134
Advances in Polymer Science

Springer

Berlin
Heidelberg
New York
Barcelona
Budapest
Hong Kong
London
Milan
Paris
Santa Clara
Singapore
Tokyo

Neutron Spin Echo Spectroscopy
Viscoelasticity
Rheology

With contributions by
B. Ewen, M. Mours, D. Richter, T. Shiga, H. H. Winter

Springer

This series presents critical reviews of the present and future trends in polymer and biopolymer science including chemistry, physical chemistry, physics and materials science. It is addressed to all scientists at universities and in industry who wish to keep abreast of advances in the topics covered.

As a rule, contributions are specially commissioned. The editors and publishers will, however, always be pleased to receive suggestions and supplementary information. Papers are accepted for „Advances in Polymer Science" in English.

In references Advances in Polymer Science is abbreviated Adv. Polym. Sci. and is cited as a journal.

Springer WWW home page: http://www.springer.de

ISSN 0065-3195
ISBN 3-540-62713-8
Springer-Verlag Berlin Heidelberg NewYork

Library of Congress Catalog Card Number 61642

Typesetting: Macmillan India Ltd., Bangalore-25
Cover: E. Kirchner, Heidelberg
SPIN: 10572847   02/3020 - 5 4 3 2 1 0 - Printed on acid-free paper

Editors

Prof. Akihiro Abe
Department of Industrial Chemistry
Tokyo Institute of Polytechnics
1583 Iiyama, Atsugi 243-02, Japan

Prof. Hans-Joachim Cantow
Freiburger Materialforschungszentrum
Stefan Meier-Str. 31a
D-79104 Freiburg i. Br., FRG

Prof. Paolo Corradini
Università di Napoli
Dipartimento di Chimica
Via Mezzocannone 4
80134 Napoli, Italy

Prof. Karel Dušek
Institute of Macromolecular Chemistry, Czech
Academy of Sciences
16206 Prague 616, Czech Republic
E-mail: OFFICE@IMC.CAS.CZ

Prof. Sam Edwards
University of Cambridge
Department of Physics
Cavendish Laboratory
Madingley Road
Cambridge CB3 OHE, UK

Prof. Hiroshi Fujita
35 Shimotakedono-cho
Shichiku, Kita-ku
Kyoto 603, Japan

Prof. Gottfried Glöckner
Technische Universität Dresden
Sektion Chemie
Mommsenstr. 13
D-01069 Dresden, FRG

Prof. Dr. Hartwig Höcker
Lehrstuhl für Textilchemie
und Makromolekulare Chemie
RWTH Aachen
Veltmanplatz 8
D-52062 Aachen, FRG

Prof. Hans-Heinrich Hörhold
Friedrich-Schiller-Universität Jena
Institut für Organische
und Makromolekulare Chemie
Lehrstuhl Organische Polymerchemie
Humboldtstr. 10
D-07743 Jena, FRG

Prof. Hans-Henning Kausch
Laboratoire de Polymères
Ecole Polytechnique Fédérale
de Lausanne, MX-D
CH-1015 Lausanne, Switzerland
E-mail: hans-henning.kausch@lp.dmx.epfl.ch

Prof. Joseph P. Kennedy
Institute of Polymer Science
The University of Akron
Akron, Ohio 44 325, USA

Prof. Jack L. Koenig
Department of Macromolecular Science
Case Western Reserve University
School of Engineering
Cleveland, OH 44106, USA

Prof. Anthony Ledwith
Pilkington Brothers plc. R & D
Laboratories, Lathom Ormskirk
Lancashire L40 SUF, UK

Prof. J. E. McGrath

Polymer Materials and Interfaces Lab.
Virginia Polytechnic and State University
Blacksburg
Virginia 24061, USA

Prof. Lucien Monnerie

Ecole Superieure de Physique et de Chimie
Industrielles
Laboratoire de Physico-Chimie
Structurale et Macromoléculaire
10, rue Vauquelin
75231 Paris Cedex 05, France

Prof. Seizo Okamura

No. 24, Minamigoshi-Machi Okazaki
Sakyo-Ku, Kyoto 606, Japan

Prof. Charles G. Overberger

Department of Chemistry
The University of Michigan
Ann Arbor, Michigan 48109, USA

Prof. Helmut Ringsdorf

Institut für Organische Chemie
Johannes-Gutenberg-Universität
J.-J.-Becher Weg 18-20
D-55128 Mainz, FRG
E-mail: Ringsdorf@mzdmza.zdv.uni-mainz.de

Prof. Takeo Saegusa

KRI International
Inc. Kyoto Research Park 17
Chudoji Minamima-chi
Shimogyo-ku Kyoto 600, Japan

Prof. J. C. Salamone

University of Lowell
Department of Chemistry
College of Pure and Applied Science
One University Avenue
Lowell, MA 01854, USA

Prof. John L. Schrag

University of Wisconsin
Department of Chemistry
1101 University Avenue
Madison, Wisconsin 53706, USA

Prof. G. Wegner

Max-Planck-Institut für Polymerforschung
Ackermannweg 10
Postfach 3148
D-55128 Mainz, FRG

Table of Contents

Neutron Spin Echo Investigations on the Segmental Dynamics of Polymers in Melts, Networks and Solutions

B. Ewen[1], D. Richter[2]
[1] MPI für Polymerforschung, Postfach 3148, 55021 Mainz, Germany
[2] Institut für Festkörperforschung, Forschungszentrum Jülich, Postfach 1913, 52425 Jülich, Germany

Neutron spin echo (NSE) spectroscopy, an advanced high-resolution quasi-elastic neutron scattering technique, provides the unique opportunity to investigate long-range relaxation processes of macromolecules simultaneously in space and time on nano-scales. In particular, information on the single-chain behavior is not restricted to dilute solutions, but may also be obtained from concentrated solutions and melts, if labelling by proton deuterium exchange is used. Thus, this method facilitates a direct microscopic study of molecular models developed to explain the macroscopic dynamic properties of polymers, e.g. transport and viscoelastic phenomena.

This article gives a short outline of the method and reviews the relevant experimental results obtained from polymer melts and networks and from dilute and semi-dilute solutions of chain molecules with different architectures since the first successful NSE work on polymers was published in 1978. The experimental observations are compared with the predictions of the related microscopic models and other theoretical approaches, which are briefly introduced and adapted accordingly.

Advances in Polymer Science, Vol. 134
© Springer-Verlag Berlin Heidelberg 1997

1 Introduction

Viscoelastic and transport properties of polymers in the liquid (solution, melt) or liquid-like (rubber) state determine their processing and application to a large extent and are of basic physical interest [1–3]. An understanding of these dynamic properties at a molecular level, therefore, is of great importance. However, this understanding is complicated by the facts that different motional processes may occur on different length scales and that the dynamics are governed by universal chain properties as well as by the special chemical structure of the monomer units [4, 5].

The earliest and simplest approach in this direction starts from Langevin equations with solutions comprising a spectrum of relaxation modes [1–4]. Special features are the incorporation of entropic forces (Rouse model, [6]) which relax fluctuations of reduced entropy, and of hydrodynamic interactions (Zimm model, [7]) which couple segmental motions via long-range backflow fields in polymer solutions, and the inclusion of topological constraints or entanglements (reptation or tube model, [8–10]) which are mutually imposed within a dense ensemble of chains.

Another approach, neglecting the details of the chemical structure and concentrating on the universal elements of chain relaxation, is based on dynamic scaling considerations [4, 11]. In particular in polymer solutions, this approach offers an elegant tool to specify the general trends of polymer dynamics, although it suffers from the lack of a molecular interpretation.

A real test of these theoretical approaches requires microscopic methods, which simultaneously give direct access to the space and time evolution of the segmental diffusion. Here, quasi-elastic scattering methods play a crucial role since they allow the measurement of the corresponding correlation functions. In particular, the high-resolution neutron spin echo (NSE) spectroscopy [12–15] is very suitable for such investigations since this method covers an appropriate range in time ($0.005 \leqslant t/ns \leqslant 40$) and space ($r/nm \lesssim 15$). Furthermore, the possibility of labelling by hydrogen-deuterium exchange allows the observation of single-chain behavior even in the melt.

This paper reviews NSE measurements on polymer melts, networks and solutions, published since the first successful NSE study on polymers [16] was performed in 1978. The experimental observations are discussed in the framework of related microscopic models, scaling predictions or other theoretical approaches.

The paper is organized in the following way: In Section 2, the principles of quasi-elastic neutron scattering are introduced, and the method of NSE is shortly outlined. Section 3 deals with the polymer dynamics in dense environments, addressing in particular the influence and origin of entanglements. In Section 4, polymer networks are treated. Section 5 reports on the dynamics of linear homo- and block copolymers, of cyclic and star-shaped polymers in dilute and semi-dilute solutions, respectively. Finally, Section 6 summarizes the conclusions and gives an outlook.

2 Quasi-Elastic Neutron Scattering

2.1 Principles of the Method

The theory and application of neutron scattering have been treated extensively in numerous monographs [17–21], different reviews [22–26] and lexicographical survey articles [27, 28]. In addition, two monographs [29, 30] deal only with quasi-elastic neutron scattering. More recently, a special monograph entitled *Polymers and Neutron Scattering* [31] has become available. Thus, here only some of the general principles of the method, being important for further understanding, are outlined.

An inelastic scattering event of a neutron is characterized by the transfer of momentum $\hbar \underline{Q} = \hbar(\underline{k}_f - \underline{k}_i)$ and energy $\hbar\omega = (\hbar^2/2m)(k_f^2 - k_i^2)$ during scattering, where $\hbar$ is the Planck constant $h/2\pi$; $\underline{k}_f$ and $\underline{k}_i$ are the final and initial wave vectors of the neutron and m is the neutron mass. For a quasi-elastic scattering process, $|\underline{k}_f| \cong |\underline{k}_i| = 2\pi/\lambda_0$ (λ_0 wavelength of the incoming neutrons) is valid. Accordingly, the magnitude of the scattering vector $\underline{Q}$ is given by

$$Q = |\underline{Q}| = \frac{4\pi}{\lambda_0} \sin \Theta \tag{1}$$

where 2Θ is the scattering angle.

The intensity of the scattered neutrons is given by the double-differential cross section $\partial^2\sigma/\partial\Omega\,\partial E$, which is the probability that neutrons are scattered into a solid angle $\partial\Omega$ with an energy change $\partial E = \hbar\omega$.

For a system containing N atoms the double-differential cross section is given by

$$\frac{\partial^2\sigma}{\partial\Omega\partial E} = \frac{1}{2\pi\hbar} \frac{k_f}{k_i} \int_{-\infty}^{+\infty} dt \exp(-i\omega t) \left\langle \sum_{j,k}^{N} b_j b_k \exp\{i\underline{Q}(\underline{r}_j(t) - \underline{r}_k(0))\} \right\rangle \tag{2}$$

where $\underline{r}_j(t)$, $\underline{r}_k(0)$ are position vectors of the atoms j and k at time t and time zero, respectively; b_j and b_k are the respective scattering lengths and $\langle \dots \rangle$ denotes the thermal average. Since neutron scattering occurs at the nuclei, the scattering lengths may depend on both the particular isotope and the relative spin orientations of neutron and nuclei. This mechanism has two consequences: (1) in general, the spin orientations of the atoms and neutrons are not correlated, giving rise to 'disorder' or incoherent scattering. This also holds for isotope distributions. (2) The fact that different isotopes of the same nucleus may have different scattering lengths can be used to label on an atomic level. In particular, the scattering lengths of H and D are significantly different ($b_H = -0.347\,10^{-12}$ cm; $b_D = 0.66\,10^{-12}$ cm), allowing easy labelling of hydrogen-containing organic matter, e.g. the conformation of a polymer in a melt can only be experimentally accessed by such a labelling technique using small-angle neutron scattering. Due to the presence of incoherent scattering, the scattering cross section (2) generally contains a coherent and an incoherent part. For simplicity considering only one type of atom, the double-differential cross section can be written as

$$\frac{\partial^2 \sigma}{\partial \Omega \partial E} = \frac{k_f}{\hbar k_i} N\{[\langle b^2 \rangle - \langle b \rangle^2] S_{inc}(\underline{Q}, \omega) + \langle b \rangle^2 S_{coh}(\underline{Q}, \omega)\}$$

with

$$S_{inc}(\underline{Q}, \omega) = \frac{1}{2\pi N} \int\limits_{-\infty}^{+\infty} e^{-i\omega t}\, dt \sum_{j,k}^{N} \langle e^{-i\underline{Q}\underline{r}_j(0)} e^{i\underline{Q}\underline{r}_k(t)} \rangle \tag{3}$$

and

$$S(\underline{Q}, \omega) \equiv S_{coh}(\underline{Q}, \omega) = \frac{1}{2\pi N} \int\limits_{-\infty}^{+\infty} e^{-i\omega t}\, dt \sum_{j,k}^{N} \langle e^{-i\underline{Q}\,\underline{r}_j(0)} e^{i\underline{Q}\,\underline{r}_k(t)} \rangle$$

where $S(\underline{Q}, \omega)$ and $S_{inc}(\underline{Q}, \omega)$ are the coherent and incoherent dynamic structure factors, respectively. $S(Q, t)$ and $S_{inc}(Q, t)$ denote the same quantities in Fourier space. The dynamic structure factors are the space-time Fourier transforms of the pair and self-correlation functions of the moving atoms. Classically, the pair-correlation function can be understood as the conditional probability of finding an atom j at location $\underline{r}$ and time t, if another atom k was at location r = 0 at time t = 0. For j = k the self-correlation function is obtained. In Gaussian approximation, i.e. assuming Gaussian distribution for the position vectors $\underline{r}_j(t)$ and $\underline{r}_k(0)$, the intermediate dynamic structure factors can be written as

$$S(Q, t) = \frac{1}{N} \sum_j \sum_k \exp\left\{ -\frac{Q^2}{6} \langle (r_j(t) - r_k(0))^2 \rangle \right\} \tag{4a}$$

$$S_{inc}(Q, t) = \frac{1}{N} \sum_j \exp\left\{ -\frac{Q^2}{6} \langle (r_j(t) - r_j(0))^2 \rangle \right\} \tag{4b}$$

Note that in this approximation the incoherent scattering measures the time-dependent thermally averaged, mean square displacement $\langle (r_j(t) - r_j(0))^2 \rangle$.

Considering only scattering at small momentum transfers, we may neglect the detailed atomic arrangements within e.g. a monomer or a solvent molecule and take into account only the scattering length densities ρ_M or ρ_S of such scattering units

$$\rho_M = \frac{\sum_j b_M^j}{v_M}; \quad \rho_S = \frac{\sum_j b_S^j}{v_S} \tag{5}$$

where the summation includes the scattering length of all atoms within a monomer or a solvent molecule; v_M and v_S are the respective molecular volumes. The scattering contrast in neutron-scattering experiments on polymers in general arises from the different scattering length densities of protonated and deuterated molecules, e.g. protonated polymers in a deuterated solvent or a deuterated polymer matrix. The contrast factor is thus defined as

$$K = (\rho^H - \rho^D) \qquad (6)$$

where ρ^H and ρ^D denote the protonated and deuterated monomers or solvent molecules respectively.

For non-interacting, incompressible polymer systems the dynamic structure factors of Eq. (3) may be significantly simplified. The sums, which in Eq. (3) have to be carried out over all atoms or in the small Q limit over all monomers and solvent molecules in the sample, may be restricted to only one average chain yielding so-called form factors. With the exception of semi-dilute solutions in the following, we will always use this restriction. Thus, S(Q, t) and S_{inc}(Q, t) will be understood as dynamic structure factors of single chains. Under these circumstances the normalized, so-called macroscopic coherent cross section (scattering per unit volume) follows as

$$\frac{\partial \Sigma}{\partial \Omega \partial E} = \phi(1 - \phi) \frac{K^2}{\hbar} \frac{V_M}{N_A} S(Q, \omega) \qquad (7)$$

with V_M the molar monomeric volume, N_A the Avogadro number and ϕ the volume fraction of the labelled polymer.

2.2 Neutron Spin Echo Technique

In this review we consider large-scale polymer motions which naturally occur on mesoscopic time scales. In order to access such times by neutron scattering a very high resolution technique is needed in order to obtain times of several tens of nanoseconds. Such a technique is neutron spin echo (NSE), which can directly measure energy changes in the neutron during scattering [32, 33].

This distinguishes NSE from conventional scattering techniques which proceed in two steps: first, monochromatization of the incident beam and then analysis of the scattered beam. Energy and momentum changes during scattering are determined by taking corresponding differences from two measurements. In order to achieve high-energy resolution with these conventional techniques, a very narrow energy interval must be selected from the relatively low intensity neutron spectrum of the source. Conventional high-resolution techniques therefore always have difficulties with low intensity.

Unlike these conventional techniques, NSE measures the neutron velocities of the incident and scattered neutrons using the Larmor precession of the neutron spin in an external magnetic field, whereby, the neutron spin vector acts like the hand of an internal clock, which is linked to each neutron and stores the result of the velocity measurement at the neutron itself. The velocity measurement is thus performed individually for each neutron. For this reason, the velocities prior to and after scattering on one and the same neutron can be directly compared and a measurement of the velocity difference during scattering becomes possible. The energy resolution is thus decoupled from the monochromatization of the incident beam. Energy resolutions of the order of 10^{-5} can be achieved with an incident neutron spectrum of 20% bandwidth.

2.2.1 Neutron Spin Manipulations with Magnetic Fields

The motion of the neutron polarization $\underline{P}(t)$ – the quantum mechanical expectancy value of neutron spin – is described by the Bloch equation

$$\frac{d}{dt}\underline{P} = \frac{2|\gamma|\mu}{\hbar}(\underline{H} \times \underline{P}) \tag{8}$$

where γ is the gyromagnetic ratio ($\gamma = -1.91$), μ the nuclear magneton and $\underline{H}$ the magnetic field. Equation (8) is the basis for manipulation of neutron polarization by external fields. Let us discuss two simple spin-angular operations. We consider a neutron beam which propagates with a polarization parallel to the direction of propagation in the z direction. A magnetic guide field parallel to z stabilizes the polarization. First, we will explain the so-called π-coil which reverses two components of the neutron spin. Its principle is shown in Figs 1a,b. A flat, long, rectangular coil, a so-called Mezei coil, is slightly tilted with respect to the x,y plane. A field H_c is generated so that the resultant field $H_\pi = H_0 + H_c$ points in the x direction. A neutron spin entering this field begins to rotate around the x-axis. During a time t = d/v, d being the coil thickness and v the neutron velocity, a phase angle $\varphi = \omega_L t$ is passed through. With the Larmor frequency $\omega_L = [(2|\gamma|\mu)/\hbar]H_\pi$ and $v = h/(m\lambda)$, we find

$$\varphi = \left(\frac{2|\gamma|\mu m 2\pi}{h^2}\right)\lambda\, dH_\pi \tag{9}$$

where φ is thus given by the line integral $\int Hds$ and is proportional to the wavelength. If we take, for example, a coil thickness of d = 1 cm and a neutron wavelength of $\lambda = 8$ Å, a field H_π of 8.5 Oe is needed to rotate the neutron spin by 180°. Such a spin-rotation operation is shown schematically in Fig. 1b. Obviously, the components of polarization x, y, z are transferred to x, − y, − z.

The second important spin-angular operation is the 90° rotation where the polarization is transformed from the z to the x direction or vice versa. A Mezei coil in the x, z plane is adjusted such that the resultant field points exactly in the direction of the bisection of the angle between x and z. A 180° rotation around this axis transforms the z component of polarization to the x direction. At the same time, the sign of the y component is inverted (Fig. 1c).

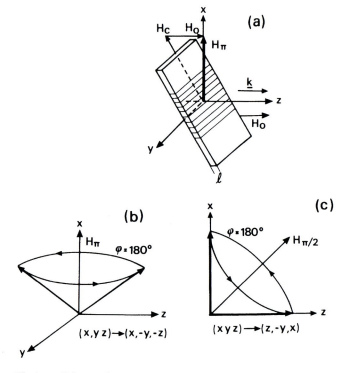

Fig. 1a–c. Spin angular operations in neutron spin-echo technique **a** Arrangement of a Mezei coil for a rotation of the neutron spin by the angle π; **b** motion of neutron polarization during the π angular operation; **c** motion of neutron polarization during a $\pi/2$ rotation. (Reprinted with permission from [12]. Copyright 1987 Vieweg and Sohn Verlagsgemeinschaft, Wiesbaden)

2.2.2 The Spin Echo Spectrometer

The basic experimental setup of a neutron spin echo spectrometer is shown in Fig. 2. A velocity selector in the primary neutron beam selects a wavelength interval of about 20% width. The spectrometer offers a primary and a secondary neutron flight path passing through the precession fields H and H'. Before beginning the first flight path, the neutron beam is polarized in the forward direction with the aid of a supermirror polarizer. A first $\pi/2$ coil rotates the polarization in the x direction perpendicular to the direction of propagation. Beginning with this well-defined, initial condition, the neutrons precess in the precession fields. Without the effect of the π-coil each neutron would go through a phase angle $\varphi \propto \lambda \int Hds$. Since the wavelengths of the neutrons are distributed over a wide range, the phase angle for each neutron would be different in front of the second $\pi/2$ coil and the beam would be completely depolarized. The π-coil, which is exactly positioned at the value of half the field integral, prevents this effect. Let the neutron spin pass through the phase angle $\varphi = n2\pi + \Delta\varphi$ on its

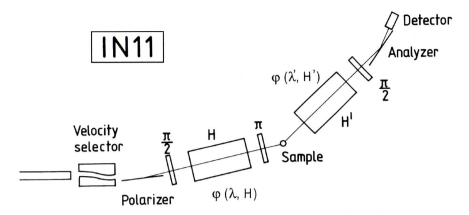

Fig. 2. Schematic representation of a neutron spin-echo spectrometer. (Reprinted with permission from [12]. Copyright 1987 Vieweg and Sohn Verlagsgemeinschaft, Wiesbaden)

way to the π-coil. The effect of the π-coil transforms the angle $\Delta\varphi_1$ to $-\Delta\varphi_1$. In a symmetric arrangement (both field integrals in front of and behind the π-coil are identical), the neutron spin passes through a second phase angle $\varphi_2 = \varphi_1 = 2n\pi + \Delta\varphi_1$. The spin transformation at the π-coil thus just compensates the angles $\Delta\varphi_1$, and in front of the second $\pi/2$ coil the neutron spin points again in the x direction irrespective of the velocity. This effect is called spin focusing or spin echo.

The second $\pi/2$ coil projects the x component of polarization in the z direction. It is then analyzed by a subsequent supermirror analyzer and detector. In a spin-echo experiment, the specimen is positioned near the π-coil. With the exception of losses due to field inhomogeneities, the polarization is maintained in the case of elastic scattering. If, however, the neutron energy is changed due to inelastic scattering processes in the specimen, then the neutron wavelength is modified from λ to $\lambda' = \lambda + \delta\lambda$. In this case, the phase angles φ_1 and φ_2 are no longer compensated. The second $\pi/2$ coil now only projects the x component of polarization from a general direction in z direction. This part of the polarization is then identified by the analyzer. Apart from resolution corrections, the final polarization P_f is then obtained from the initial polarization P_i as

$$P_f = P_i \int_{-\infty}^{+\infty} S(Q,\omega)\cos\omega t \, d\omega \bigg/ \int_{-\infty}^{+\infty} S(Q,\omega) \, d\omega \qquad (10)$$

The normalized dynamic structure factor thus gives the probability that a scattering event occurs at a certain wavelength change $\delta\lambda = (m/2\pi)\lambda^3\omega$ at a given momentum transfer Q. The Fourier time t in the argument of the cosine is determined by the transformation from the phase angle $\Delta\varphi = \varphi_2 - \varphi_1$ to ω. The numerical relation reads

$$t[s] = 1.863 \times 10^{-14} H L \lambda^3 \qquad (11)$$

The length L of the precession field is measured in meters, the guide field H in Oersted and the wavelength λ in Ångstrøm.

Obviously, the final polarization P_f is directly connected with the normalized intermediate dynamic structure factor $S(Q, t)/S(Q)$. In order to estimate typical Fourier times accessed by this method we consider a guide field of 500 Oe in a precession coil of $L = 200$ cm operating at a neutron wavelength of 8 Å. Then a time of about 10 ns is reached. A spin-echo scan is performed by varying the guide field and studying the intermediate scattering function $S(Q, t)$ at different Fourier times. Finally, we note that the use of a broad wavelength band introduces a further averaging process containing an integration over the incident wavelength distribution $I(\lambda)$,

$$P_f = P_i \int_0^\infty \frac{I(\lambda)S[Q(\lambda), t(\lambda)]\, d\lambda}{S(Q(\lambda))\, d\lambda} \tag{12}$$

This averaging process somewhat obscurses the relationship between P_f and $S(Q, t)$. For many relaxation processes, however, where the quasi-elastic width varies with power laws in Q, the smearing of (12) is of no practical importance. For example, for internal relaxation of polymers in dilute solution, we have $S(Q, t) = S(Q^2 t^{2/3})$ [34]. Since Q varies with $1/\lambda$ and t with λ^3, the wavelength dependence drops out completely. For illustration, Fig. 3 shows a technical

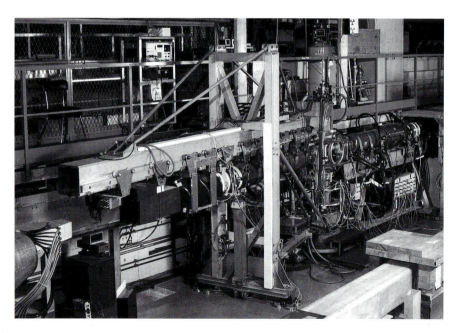

Fig. 3. The neutron spin echo spectrometer IN11 at the ILL Grenoble. The two large coils producing the precession fields are clearly visible (Photograph was kindly given to us by H. Büttner, Scientific Coordination Office of the Institute Laue-Langevin Grenoble).

realization of the spin-echo spectrometer at the Institute Laue-Langevin, Grenoble. Particularly striking are the Larmor precession coils in front and behind the specimen. Furthermore, some coils causing the spin-angular operations can be recognized.

3 Polymer Motion in Dense Environments

Long-chain linear polymers in the melt or in concentrated solution show a number of anomalous viscoelastic properties [1–3, 5]. Particularly spectacular is the appearance of a plateau in the time dependence of the dynamic modulus, which expands more and more with growing chain length. In this plateau region, stress is proportional to strain whereby Hooke's law applies: although a liquid, the polymer melt behaves elastically. This behavior reminds us of the rubber elasticity of chemically cross-linked polymers. However, the modulus appearing there does not decrease to zero over long times, since the slipping of neighbored chains is impossible due to the existence of permanent cross-links. The magnitude of the modulus is proportional to the temperature and inversely proportional to the mesh size of the network. The proportionality to temperature is caused by entropic forces resulting from the conformational entropy of the coiled chains between the network points. This conformational entropy follows from the number of possible arrangements of chain elements in space. According to the central limit theorem, the most probable arrangement is that of a Gaussian coil, i.e. the polymer chain performs a random walk in space. If part of the chain is stretched, an entropic force resulting from the reduction of free energy acts with the aim of restoring the more probable coiled state for these stretched segments.

Based on these arguments it is appropriate to assume that the elastic behavior of polymer melts is caused by network-like entanglements of the polymer chains, so that the melt becomes a temporary network. The role of the network points is taken over by entanglements. The plateau modulus is ascribed to the rubber elastic modulus of this temporary network. On this assumption it is possible to estimate the mean distance between the hypothetical entanglement points in the melt from the value of the plateau modulus. The resultant entanglement distances d are typically of the order of $d = 40$–$100\,\text{Å}$. Compared with the two characteristic length scales of a polymer, the monomer length $\sigma \sim 5\,\text{Å}$ and the end-to-end distance $R_e \sim 1000\,\text{Å}$ of the coiled polymer chain, the entanglement distance defines an intermediate length scale of a dynamic nature.

The different length scales involve different time scales with different types of motion. For short times corresponding to spatial distances shorter than the entanglement distance, we expect entropy-determined dynamics described by the so-called Rouse model [6, 35.]. As the spatial extent of motion increases and reaches dimensions of the entanglement distance, the freedom of motion of the chain is greatly restricted. The temporary network leads to a localization of the chain which can only move like a snake along its own profile due to the network which restricts lateral motions. For very long times, the chain leaves its original topological restrictions behind – the plateau modulus relaxes. This concept is essentially contained in the reptation model of de Gennes [8, 10]. The length and time scales of importance for viscoelasticity start at the bond length in the region of 1.5 Å and extend up to the chain dimension in the region of 1000 Å. The associated times start in the picosecond range and can reach macroscopic dimensions.

This section presents results of the space-time analysis of the above-mentioned motional processes as obtained by the neutron spin echo technique. First, the entropically determined relaxation processes, as described by the Rouse model, will be discussed. We will then examine how topological restrictions are noticed if the chain length is increased. Subsequently, we address the dynamics of highly entangled systems and, finally, we consider the origin of the entanglements.

3.1. Rouse Dynamics

3.1.1. Theoretical Outline – the Rouse Model

We consider a Gaussian chain in a heat bath as the simplest molecular model (Rouse model, [6]). The building blocks of such a chain are statistical segments which consist of several main chain bonds such that its end-to-end distance follows a Gaussian distribution. The whole chain is then a succession of freely joined Gaussian segments of a length ℓ. Neglected are excluded volume effects and all hydrodynamic interactions – both are screened in a dense environment and will be discussed in the section on polymer solutions – as well as the topological interactions of the chains, which are the subject of the reptation model discussed later. Under these circumstances the Brownian motion of the chain segments results from entropic forces, which arise from the conformational entropy of the chain, and stochastic forces from the heat bath. The Langevin equation for segment 'n' follows:

$$\zeta \frac{d\underline{r}_n}{dt} = \nabla_n F(\underline{r}_n) + \underline{\ell}_n(t) \tag{13}$$

where ζ is the monomeric friction coefficient and $\underline{r}_n$ the position vector of segment 'n'. For the stochastic force $\underline{\ell}_n(t)$, we have $\langle \underline{\ell}_n(t) \rangle = 0$ and $\langle f_{n\alpha}(t) f_{m\beta}(t') \rangle = 2k_B T \zeta \delta_{nm} \delta_{\alpha\beta} \delta(t - t')$ (white noise), α and β denote the Cartesian components of $\underline{r}$. $F(\underline{r}_n)$ is the free energy of the polymer chain. The force term in Eq. (13) is dominated by the conformational entropy of the chain $S = k_B \ln W(\{\underline{r}_n\})$ (k_B Boltzmann constant). The probability $W(\{\underline{r}_n\})$ for a chain conformation of a Gaussian chain of N segments is

$$W(\{\underline{r}_n\}) = \prod_{i=1}^{N} \left(\frac{3}{2\pi\ell^2}\right)^{3/2} \exp\left(\frac{-3(\underline{r}_i - \underline{r}_{i-1})^2}{2\ell^2}\right) \tag{14}$$

We note that the resulting entropic force is particular to macromolecular systems, where the conformational entropy generates a force stabilizing the most favorable conformation.

With the boundary conditions that the chain ends are free of forces, Eq. (13) is readily solved by cos-Fourier transformation, resulting in a spectrum of normal modes. Such solutions are similar, e.g. to the transverse vibrational modes of a linear chain except that relaxation motions are involved here instead of periodic vibrations.

We obtain for the relaxation rates

$$\tau_p^{-1} = W\left(1 - \cos\frac{\pi p}{N}\right) \quad p = 1, 2, \dots, N \tag{15}$$

where

$$W = \frac{3k_BT}{\zeta\ell^2} \tag{16}$$

and the dimension of reciprocal time is the elementary relaxation rate of the Rouse process.

The mode index p counts the number of modes along the chain. A small mode number, e.g. $p \ll N$, (15), is approximated by

$$\tau_p^{-1} = \frac{3\pi^2 k_BTp^2}{\zeta N^2\ell^2} = \frac{p^2}{\tau_R} \quad p = 1, 2, \dots, N \tag{17}$$

where τ_R, the longest time of the relaxation spectrum, is also called the Rouse time of the chain. For the correlation function in Fourier space,

$$\langle x_p^\alpha(t)x_p^\beta(0)\rangle = \delta_{\alpha\beta}\delta_{pq}\frac{N\ell^2}{6\pi^2p^2}\exp(-t/\tau_p)$$

$$\langle x_0^\alpha(t)x_0^\beta(0)\rangle = \delta_{\alpha\beta}\frac{2kT}{N\zeta}t \tag{18}$$

is obtained, where x_p^α is the α component of the number 'p' normal mode and x_0^α the related center of the mass coordinate. In order to study the Brownian motion, the segment correlation functions in real space $\Delta r_{nm}^2(t) = \langle(r_m(t) - r_n(0))^2\rangle$ are significant. They are obtained by retransformation of the normal coordinates leading to

$$\Delta r_{mn}^2(t) = 6D_Rt + |n - m|\ell^2$$

$$+ \frac{4N\ell^2}{\pi^2}\sum_{p=1}^{N}\frac{1}{p^2}\cos\left(\frac{p\pi m}{N}\right)\cos\left(\frac{p\pi n}{N}\right)\left(1 - \exp\left(-\frac{p^2t}{\tau_R}\right)\right)$$

$$\tag{19}$$

In Eq. (19) we used the fact that the mean square displacement of the center-of-mass provides the diffusion constant according to $D_R = (1/6t)\langle(\underline{x}_0(t) - \underline{x}_0(0))^2\rangle$. For the special case of the self correlation function $(n = m)\,\Delta r_{nm}(t)$ reveals the mean square displacement of a polymer segment. For $t < \tau_R$ we obtain

$$\Delta r_{nn}^2 = 2\ell^2 \left(\frac{3k_B\,Tt}{\pi\zeta\ell^2}\right)^{1/2} \tag{20}$$

In contrast to normal diffusion, Δr_{nn}^2 does not grow linearly but with the square root of time. This may be considered the result of superimposing two random walks. The segment executes a random walk on the random walk given by the chain conformation. For the translational diffusion coefficient $D_R = k_B T/\,N\zeta$ is obtained: D_R is inversely proportional to the number of friction-performing segments.

Dynamic Structure Factors for the Rouse Model [35]

The prerequisite for an experimental test of a molecular model by quasi-elastic neutron scattering is the calculation of the dynamic structure factors resulting from it. As outlined in Section 2 two different correlation functions may be determined by means of neutron scattering. In the case of coherent scattering, all partial waves emanating from different scattering centers are capable of interference; the Fourier transform of the pair-correlation function is measured Eq. (4a). In contrast, incoherent scattering, where the interferences from partial waves of different scatterers are destructive, measures the self-correlation function [Eq. (4b)].

    The self-correlation function leads directly to the mean square displacement of the diffusing segments $\Delta r_{nn}^2(t) = \langle(r_n(t) - r_n(0))^2\rangle$. Inserting Eq. (20) into the expression for $S_{inc}(Q, t)$ [Eq. (4b)] the incoherent dynamic structure factor is obtained

$$\frac{S_{inc}(Q, t)}{S_{inc}(Q, 0)} = \exp - \left\{\frac{2}{\sqrt{\pi}}(\Omega_R(Q)t)^{1/2}\right\} \tag{21}$$

with the characteristic rate of the Rouse dynamics

$$\Omega_R(Q) = \frac{1}{12}k_B T\frac{\ell^2}{\zeta}\,Q^4 \tag{22}$$

Unlike for diffusion, where the characteristic relaxation rate is proportional to Q^2, here the fourth power of Q is found.

    It is obvious that (21) is equivalent to a stretched exponential decay function of the general form

$$S_{inc}(Q, t) \sim \exp - \left\{\frac{Q^2\ell^2}{6}(t/\tau)^\beta\right\} \tag{23}$$

where the line shape parameter β has the value $1/2$.

In the case of coherent scattering, which observes the pair-correlation function, interference from scattering waves emanating from various segments complicates the scattering function. Here, we shall explicitly calculate $S(Q, t)$ for the Rouse model for the limiting cases (1) $QR_e \ll 1$ and (2) $QR_e \gg 1$ where $R_e^2 = \ell^2 N$ is the end-to-end distance of the polymer chain.

By substituting Eq. (19) in (4a) we obtain for the Rouse model

$$S(Q, t) = \frac{1}{N} \sum_{n, m} \exp \left\{ - Q^2 D_R t - \frac{1}{6} |n - m| Q^2 \ell^2 \right.$$

$$\left. - \frac{2N\ell^2 Q^2}{3\pi^2} \sum_p \frac{1}{p^2} \cos \left(\frac{p\pi n}{N} \right) \cos \left(\frac{p\pi m}{N} \right) \left(1 - \exp \left(- \frac{tp^2}{\tau_R} \right) \right) \right\}$$

(24)

(1) For small angles ($QR_e \ll 1$), the second and third terms in (24) are negligibly small and $S(Q, t)$ describes the center-of-mass diffusion of the coil

$$S(Q, t) = N \exp \left(- Q^2 \frac{k_B T}{N\zeta} t \right)$$

(25)

(2) For $QR_e \gg 1$ we shall restrict ourselves to the internal relaxation $t < \tau_R$. We thus transform Eq. (19)

$$\Delta r_{nm}^2(t) = 6 D_R t + |n - m| \ell^2 + \frac{2N\ell^2}{\pi^2} \sum_p \frac{1}{p^2} \left\{ \cos \left(\frac{p\pi(n + m)}{N} \right) \right.$$

$$\left. + \cos \left(\frac{p\pi(n - m)}{N} \right) \left(1 - \exp - \frac{tp^2}{\tau_R} \right) \right\}$$

(26)

The total is dominated by large p values for $t \gg \tau_R$. The underlined cos-term then oscillates very rapidly and its contribution can be neglected. We substitute the remainder in (4a) and convert the p summation into an integral

$$S(Q, t) = \frac{1}{N} \int_0^N dn \int_0^N dm \exp \left[- \frac{1}{6} Q^2 |n - m| \ell^2 \right.$$

$$\left. - \frac{Q^2 N \ell^2}{3\pi^2} \int_0^\infty dp \frac{1}{p^2} \cos \left(\frac{p\pi(n - m)}{N} \right) \left(1 - \exp \left(- \frac{tp^2}{\pi_R} \right) \right) \right]$$

(27)

For further calculations, the following substitutions are applied
1. $\int dn \int dm \rightarrow \int_{-\infty}^{+\infty} d(n - m)$ (expansion of the limits to infinity is permitted since the main contribution of the integral is obtained for $n \cong m$),
2. $|n - m| \rightarrow 6u/q^2\ell^2$, $p^2 \rightarrow (\zeta N^2 \ell^2 / 3\pi^2 k_B T t) x^2$.

Table 1. Dynamic structure factors for Rouse and Zimm dynamics

	Coherent dynamic structure factor	Characteristic frequency	Adjustable parameter
Rouse dynamics [35] $B = 0$	$\dfrac{S(Q,t)}{S(Q,0)} = \int_0^\infty dy\, e^{-u} \exp\left\{-(\Omega_R t)^{1/2} h(u(\Omega_R t)^{-1/2})\right\}$ $h(y) = \dfrac{2}{\pi}\int_0^\infty dx\, \dfrac{\cos(xy)}{x^2}(1-\exp(-x^2))$	$\Omega_R = \dfrac{1}{12}\dfrac{k_B T}{\zeta}\ell^2 Q^4$	$\dfrac{\ell^2}{\zeta}$
Zimm dynamics [34] $B \simeq 0.4$	$\dfrac{S(Q,t)}{S(Q,0)} = \int_0^\infty dy\, e^{-y} \exp\left\{-(\Omega_Z t)^{2/3} h(y(\Omega_Z t)^{-2/3})\right\}$ $h(u) = \dfrac{2}{\pi}\int_0^\infty dx\, \dfrac{\cos(xu)}{x^2}\left\{1-\exp(-2^{-1/2}x^{3/2})\right\}$	$\Omega_Z = \dfrac{1}{6\pi}\dfrac{k_B T}{\eta_s} Q^3$	none
Combined Rouse and Zimm dynamics [40] $0 \le B \le 0.4$	$\dfrac{S(Q,t)}{S(Q,0)} = \int_0^\infty dy\, e^{-y} \exp\left\{-f(y,\Omega t, B)\right\}$ $f(y,\Omega t, B) = \dfrac{2}{\pi}\int_0^\infty dx\, \dfrac{\cos(xy)}{x^2}(1-\exp(-\tilde{B}tx^2))$ $\tilde{B} = \dfrac{1+(B/Q\ell)(12\pi/x)^{1/2}}{1+(2B/Q\ell)(6\pi)^{1/2}}$	$\Omega = \dfrac{1}{12}\dfrac{k_B T}{\zeta}\ell^2 Q^4\left(1+2\sqrt{6\pi}\,\dfrac{B}{Q\ell}\right)$	$\dfrac{\ell^2}{\zeta}; \ell$

Regime of validity $N \to \infty$, $(N\ell^2/6)^{-1/2} \ll Q \ll \ell^{-1}$; B draning parameter [see Eq. (76)]

We obtain

$$S(Q,t) = \frac{12}{Q^2 \ell^2} \int_0^\infty du \exp\{-u - (\Omega_R t)^{1/2} h(u(\Omega_R t^{1/2}))\} \tag{28}$$

with

$$h(y) = \frac{2}{\pi} \int_0^\infty dx \frac{\cos(xy)}{x^2}(1 - \exp(-x^2)) \quad \text{(see Table 1)}.$$

We observe that in spite of the complicated functional form, $S(Q,t)$, like the self-correlation function, depends only on one variable, the Rouse variable

$$(\Omega_R t)^{1/2} = \frac{Q^2}{6}\sqrt{\frac{3k_B T \ell^2 t}{\zeta}} \equiv \frac{Q^2 \ell^2}{6}\sqrt{Wt} \tag{29}$$

For different momentum transfers the dynamic structure factors are predicted to collapse to one master curve, if they are represented as a function of the Rouse variable. This property is a consequence of the fact that the Rouse model does not contain any particular length scale. In addition, it should be mentioned that ℓ^2/ζ or the equivalent quantity $W\ell^4$ is the only adjustable parameter when Rouse dynamics are studied by NSE.

3.1.2 NSE Observations on Rouse Dynamics

From a theoretical point of view, Rouse dynamics should prevail (1) for short-chain polymer melts with molecular weights (M_W) below the entanglement limit and (2) for long chains at short enough times. If a polymer exhibits a low plateau modulus, i.e. entanglement constraints are only weakly developed, then this initial Rouse regime may be rather extended in time, facilitating a clear observation. Experimental investigations of the dynamic structure factor for both cases have been reported [36–44]. While for long-chain melts the asymptotic scattering functions of Eq. (28) apply, short-chain dynamics require the explicit consideration of single modes Eq. (24). In this section we deal with the long-chain limit, while results on short chains will be presented in connection with the entanglement transition described in the next chapter.

Poly(dimethylsiloxane) (PDMS) Melts

Rouse motion has been best documented for PDMS [38–44], a polymer with little entanglement constraints, high flexibility and low monomeric friction. For this polymer NSE experiments were carried out at $T = 100\,°C$ to study both the self- and pair-correlation function.

Measurements of the self-correlation function with neutrons are normally performed on protonated materials since incoherent scattering is particularly strong there. This is a consequence of the spin-dependent scattering lengths of hydrogen. Due to spin-flip scattering, which leads to a loss of polarization, this

approach cannot be used for the NSE method. In order to overcome this
obstacle, experiments on the self-correlation function were performed on
a deuterated PDMS sample ($M_W = 100\,000$ g/mol) containing short protonated
sequences at random intervals. Each of these protonated sequences contained
eight monomers. In such a specimen, scattering is mainly caused by the contrast
between the protonated sequence and the deuterated environment and is therefore coherent. On the other hand, the sequences are randomly distributed, so
that there is no constructive interference of partial waves produced at different
sequences. The scattering experiment measures the self-correlation function
under these conditions [41].

In Fig. 4 the scattering data are plotted versus the Rouse variable [Eq. (29)].
The results for different momentum transfers follow a common straight line. For

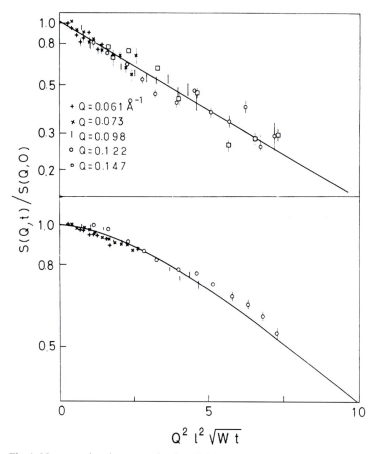

Fig. 4. Neutron spin echo spectra for the self-(*above*) and pair-(*below*) correlation functions obtained
from PDMS melts at 100 °C. The data are scaled to the Rouse variable. The *symbols* refer to the
same Q-values in both parts of the figure. The *solid lines* represent the results of a fit with the
respective dynamic structure factors. (Reprinted with permission from [41]. Copyright 1989 The
American Physical Society, Maryland)

the case of the self-correlation function, the scattering function directly measures the mean square displacement which, according to Eq. (20), follows a square root law in time. This behavior is apparent Fig. 4.

The pair-correlation function for the segmental dynamics of a chain is observed if some protonated chains are dissolved in a deuterated matrix. The scattering experiment then observes the result of the interfering partial waves originating from the different monomers of the same chain. The lower part of Fig. 4 displays results of the pair-correlation function on a PDMS melt ($M_W = 1.5 \times 10^5$, $M_W/M_W = 1.1$) containing 12% protonated polymers of the same molecular weight. Again, the data are plotted versus the Rouse variable.

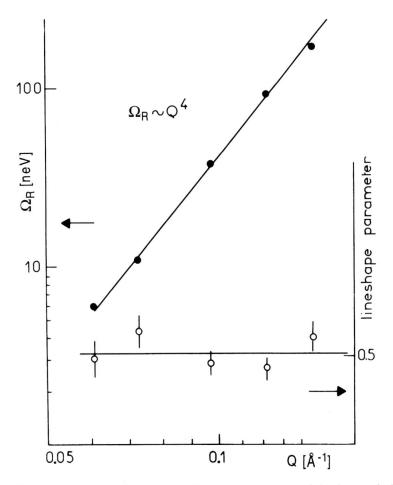

Fig. 5. The characteristic frequencies Ω_R and time exponents β in the stretched exponential relaxation function obtained for the randomly labelled PDMS melt at 100 °C. (Reprinted with permission from [44]. Copyright 1989 Steinkopff Verlag, Darmstadt)

Though the functional form of the dynamic structure factor is more complicated than that for the self-correlation function, the data again collapse on a common master curve which is described very well by Eq. (28). Obviously, this structure factor originally calculated by de Gennes, describes the neutron data well (the only parameter fit is $W\ell^4 = 3k_BT\ell^2/\zeta$) [41, 44].

Thus, excellent agreement with theoretical predictions is observed for both correlation functions. This refers to both Q and time dependence. The characteristic frequencies of both data sets follow the theoretical Q^4 law, and the line shape of the dynamic structure factor of the self-correlation function fulfills the predicted $t^{1/2}$ behavior. This is illustrated in Fig. 5 for the case of the self-correlation function. It shows the characteristic frequency Ω_R and the experimental time exponent in the stretched exponential scattering function [Eq. (23)] as a function of Q. Within the framework of the experimental accuracy, the Rouse rates determined from both measurements are identical (self-correlation: $W\ell^4 = 1.75 \pm 0.15 \times 10^{13}\text{Å}^4\text{s}^{-1}$; pair correlation $1.85 \pm 0.1 \times 10^{13}\text{Å}^4\text{s}^{-1}$). From these findings the monomeric friction coefficient per mean square segment length $\zeta/\ell^2 = 8.9 \pm 0.7 \times 10^5 \text{ dyn s/cm}^3$ is derived.

On the other hand, the viscosity η of a non-entangled melt is expressed by

$$\eta = \frac{\rho N_A}{M} \frac{k_B T}{2} \sum_p \tau_p = \frac{N_A}{36} \frac{\rho}{M_0} \zeta\ell^2 N \tag{30}$$

where M_0 is the molar monomer mass and ρ the polymer density. Thus, a direct comparison between microscopic (NSE) and macroscopic (rheological) data requires the additional knowledge of ℓ^2, which is 39.7 Å^2 in the case of PDMS. On the basis of this value, $\zeta/\ell^2 = 7.1 \ 10^5 \text{ dyn s/cm}^3$ is obtained from viscosity measurements which agrees well with the above given number, derived from NSE data [44, 45].

Poly(isoprene) (PI) Melts

As a second example we present recent NSE results on PI melts ($M_w = 5.7 \times 10^4$, $M_w/M_n = 1.02$) where again protonated molecules (10%) in a deuterated matrix were studied at T = 200 °C [39].

Figure 6 shows the measured dynamic structure factors for different momentum transfers. The solid lines display a fit with the dynamic structure factor of the Rouse model, where the time regime of the fit was restricted to the initial part. At short times the data are well represented by the solid lines, while at longer times deviations towards slower relaxations are obvious. As it will be pointed out later, this retardation results from the presence of entanglement constraints. Here, we focus on the initial decay of S(Q,t). The quality of the Rouse description of the initial decay is demonstrated in Fig. 7 where the Q-dependence of the characteristic decay rate Ω_R is displayed in a double logarithmic plot. The solid line displays the $\Omega_R \sim Q^4$ law as given by Eq. (29).

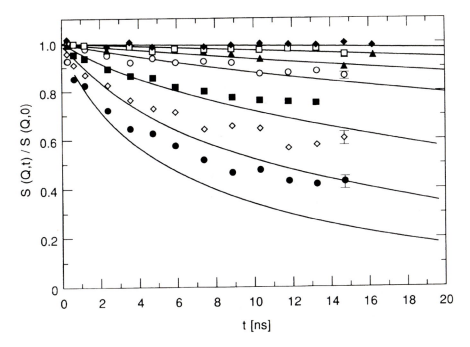

Fig. 6. Dynamic structure factor as observed from PI for different momentum transfers at 468 K. (◆ $Q = 0.038 \, \text{Å}^{-1}$; □ $Q = 0.051 \, \text{Å}^{-1}$; △ $Q = 0.064 \, \text{Å}^{-1}$; ○ $Q = 0.077 \, \text{Å}^{-1}$; ■ $Q = 0.102 \, \text{Å}^{-1}$; ◇ $Q = 0.128 \, \text{Å}^{-1}$; ● $Q = 0.153 \, \text{Å}^{-1}$). The *solid lines* display fits with the Rouse model to the initial decay. (Reprinted with permission from [39]. Copyright 1992 American Chemical Society, Washington)

Over the entire Q-range within experimental error the data points fall on the line and thus exhibit the predicted Q^4 dependence. The insert in Fig. 7 demonstrates the scaling behavior of the experimental spectra which, according to the Rouse model, are required to collapse to one master curve if they are plotted in terms of the Rouse variable $u = Q^2 \ell^2 \sqrt{Wt}$. The solid line displays the result of a joint fit to the Rouse structure factor with the only parameter fit being the Rouse rate $W\ell^4$. Excellent agreement with the theoretical prediction is observed. The resulting value is $W\ell^4 = 2.0 \pm 0.1 \times 10^{13} \, \text{Å}^4 \, \text{s}^{-1}$.

Rouse behavior observed on PI homopolymer melts has to be modified if the labelled (protonated) PI species are replaced by diblock copolymers of protonated PI and deuterated polystyrene (PS) [46]. The characteristic frequency $\Omega(Q)$ is slowed down considerably due to the presence of the non-vanishing χ-parameter. Thus, the reduction is stronger at smaller Q-values or at larger length scales than in the opposite case. In addition, as a minor effect, $\Omega(Q)$ becomes dependent on both friction coefficients per mean square monomer length, ζ/ℓ^2, valid for PI and for PS.

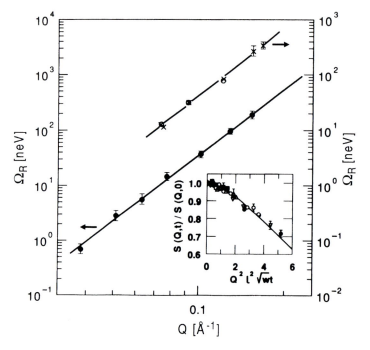

Fig. 7. Characteristic relaxation rate for the Rouse relaxation in polyisoprene as a function of momentum transfer. The *insert* shows the scaling behavior of the dynamic structure factor as a function of the Rouse variable. The different *symbols* correspond to different Q-values. (Reprinted with permission from [39]. Copyright 1992 American Chemical Society, Washington)

In summary, the chain dynamics for short times, where entanglement effects do not yet play a role, are excellently described by the picture of Langevin dynamics with entropic restoring forces. The Rouse model quantitatively describes (1) the Q-dependence of the characteristic relaxation rate, (2) the spectral form of both the self- and the pair correlation, and (3) it establishes the correct relation to the macroscopic viscosity.

3.2 Transition from Rouse to Entanglement Controlled Dynamics

Figures 8 and 9 show the dependence of the self-diffusion constant and the viscosity of polyethylene melts on molecular weight [47, 48]. For small molecular weights the diffusion constant is inversely proportional to the chain length – the number of frictional monomers grows linearly with the molecular weight. This behavior changes into a $1/M^2$ law with increasing M. The diffusion

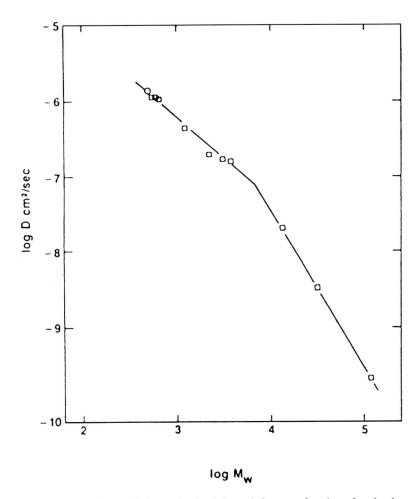

Fig. 8. Self-diffusion coefficients of polyethylene chains as a function of molecular mass. The measurements were carried out at the same value of the monomeric friction coefficient. (Reprinted with permission from [48]. Copyright 1987 American Chemical Society, Washington)

constant is much more reduced than would be expected from the increasing number of frictional monomers. This regime is described by the reptation model which will be discussed in more detail in the following section. A similar transition is found in the molecular weight dependence of viscosity. For short chains the linear relationship between viscosity and chain length predicted by Eq. (30) is observed, whereas for longer chains the viscosity depends on the molecular weight with a high power ($M^{3.4}$); this transition is again ascribed to the entanglements.

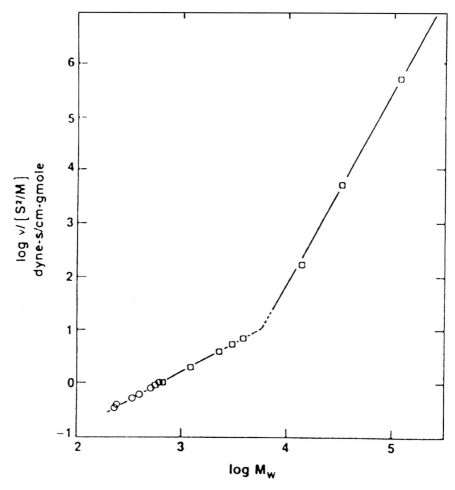

Fig. 9. Kinematic viscosity $\nu = \eta/\rho$ (η viscosity, ρ density) divided by S^2/M ($S^2 \equiv \langle R_g^2 \rangle$ mean square radius of gyration, M molecular mass) as a function of M for polyethylene melts at the same monomeric friction coefficient. (Reprinted with permission from [48]. Copyright 1987 American Chemical Society, Washington)

3.2.1 Mode Analysis and Generalized Rouse Model

For Gaussian chains the spatial structure of the eigenmodes is given by the Rouse form

$$x_p = \frac{1}{N} \sum_n r_n(t) \cos\left(\frac{pn\pi}{N}\right) \tag{31}$$

while the time dependence may have an arbitrary form. This a priori knowledge
of the spatial form of the eigenmodes is the starting point for an analysis asking
how the different normal modes of a chain are influenced if topological constraints become important [36]. Using the Rouse form of the eigenmodes and
allowing for a general time dependence, the dynamic structure factor of Eq. (24)
may be generalized:

$$S(Q, t) = \frac{1}{N} \exp\left(-\frac{Q^2}{6} \langle [x_0(t) - x_0(0)]^2 \rangle \right)$$

$$\times \left\{ \sum_{n, m} \exp\left[-\frac{1}{6} Q^2 \ell^2 |m - n| - \frac{4 R_g^2 Q^2}{\pi^2} \sum_{p=1}^{N} \frac{1}{p^2} \right. \right.$$

$$\left. \left. \times \cos\left(\frac{p\pi m}{N} \right) \cos\left(\frac{p\pi n}{N} \right) (1 - \langle x_p(t) x_p(0) \rangle) \right] \right\} \qquad (32)$$

where x_0 denotes the center of mass coordinates and the curly brackets denote
the thermal average. The correlation function $\langle (x_0(t) - x_0(0))^2 \rangle$ describes the
diffusive motion of the molecular center of gravity. The relaxation dynamics of
the internal modes are hidden in the exact time dependence of the correlators
$\langle x_p(t) x_p(0) \rangle$. They describe the time-dependent development of the motion of
a normal mode p. In the case of entropy-determined Rouse motion, the correlators have been given in Section 3.1, Eq. (18). The beginning spatial restrictions
should lead to a more complicated time dependence.

How can one hope to extract the contributions of the different normal modes
from the relaxation behavior of the dynamic structure factor? The capability of
neutron scattering to directly observe molecular motions on their natural time
and length scale enables the determination of the mode contributions to the
relaxation of S(Q, t). Different relaxation modes influence the scattering function
in different Q-ranges. Since the dynamic structure factor is not simply broken
down into a sum or product of more contributions, the Q-dependence is not
easy to represent. In order to make the effects more transparent, we consider the
maximum possible contribution of a given mode p to the relaxation of the
dynamic structure factor. This maximum contribution is reached when
the correlator $\langle x_p(t) x_p(0) \rangle$ in Eq. (32) has fallen to zero. For simplicity, we retain
all the other relaxation modes: $\langle x_s(t) x_s(0) \rangle = 1$ for $s \neq p$.

Under these conditions, Eq. (32) indicates the maximum extent to which
a particular mode p can reduce S(Q, t) as a function of the momentum transfer
Q. Figure 10 presents the Q-dependence of the mode contributions for PE of
molecular weights $M_W = 2000$ and $M_W = 4800$ used in the experiments to be
described later. Vertical lines mark the experimentally examined momentum
transfers. Let us begin with the short chain. For the smaller Q the internal modes
do not influence the dynamic structure factor. There, only the translational
diffusion is observed. With increasing Q, the first mode begins to play a role. If
Q is further increased, higher relaxation modes also begin to influence the

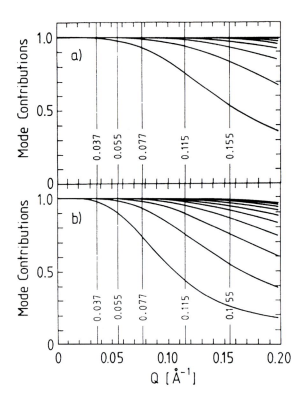

Fig. 10a, b. Contributions of the different modes to the relaxation of the dynamic structure factor $S(Q, t)/S(Q, 0)$ (see text) for PE of molecular masses. **a** $M_w = 2.0 \times 10^3$ g/mol and **b** $M_w = 4.8 \times 10^3$ g/mol. The experimental Q-values are indicated by *vertical lines; curves* correspond to mode numbers increasing from *bottom* to *top*. (Reprinted with permission from [52]. Copyright 1993 The American Physical Society, Maryland)

dynamic structure factor. If the chain length is enlarged, the influence of the different internal relaxation modes shifts toward smaller momentum transfers. This Q-dependence of the contributions of different relaxation modes to $S(Q, t)$ permits a separation of the influence of different modes on the dynamic structure factor.

In order to be able to evaluate data with a reasonable number of parameters, the mode analysis assumes, as a first approximation, that the exponential correlation of the correlations [Eq. (18)] is maintained, and only the relaxation rates $1/\tau_p$ are allowed to depend on a general form on the mode index

$$\frac{1}{\tau_p} = \frac{\pi^2 p^2}{N^2} W_p \tag{33}$$

The mode-dependent rate W_p is the parameter of analysis.

Some years ago, on the basis of the excluded-volume interaction of chains, Hess [49] presented a generalized Rouse model in order to treat consistently the dynamics of entangled polymeric liquids. The theory treats a generalized Langevin equation where the entanglement friction function appears as a kernel of a memory function

$$\frac{3kT}{\ell^2}(r_{n+1}(t) + r_{n-1}(t) - 2r_n(t)) + \zeta\dot{r}_n(t)$$

$$+ \sum_m \int_0^t dt'\, \tau\Delta\zeta_{nm}(t-t')\dot{r}_m(t') = f_n(t) \tag{34}$$

where f_n the random forces acting on segment n, and $\Delta\zeta_{nm}(t)$ is the entanglement friction function. The entanglement friction function is calculated as the auto-correlation function of the excluded-volume forces. Though important approximations are involved, this approach goes beyond earlier phenomenological attempts, where $\Delta\zeta$ was derived from phenomenological arguments [50]. In order to obtain a treatable problem, a series of approximations were necessary, reducing the many chain problem of Eq. (34) to a single-chain equation. Thus, the entanglement fricton function is expressed in the form of the single-chain propagator squared, including not only motions of single chains but also that of other chains, or constraint-release processes are built in. This approach has some similarity to des Cloiseaux's recent double reptation concept [51]. Finally, the entanglement friction function becomes

$$\Delta\zeta(t) = 4\,\frac{k_B T}{N_c}\, q_c^2 \int_0^2 dy\, y^4 \exp(-ty^2/\tau_{eff}) \tag{35}$$

where τ_{ef} is the characteristic time for the relaxation of the entanglement friction function and q_c^{-1} stands for the range of the chain interaction potential. Taking $\Delta\zeta(t=0)$ and N_c as the two model parameters τ_{eff} may be written as

$$\tau_{eff} = \frac{N}{N_c}\,\frac{\zeta}{\Delta\zeta(0)} \begin{cases} \dfrac{2}{5}\left(1 - \dfrac{2}{3}\dfrac{N}{N_c}\right)^{-1} & N < N_c \\[2ex] \dfrac{6}{5} & N > N_c \end{cases} \tag{36}$$

Equations (35) and (36) define the entanglement friction function in the generalized Rouse equation (34) which now can be solved by Fourier transformation, yielding the frequency-dependent correlators $\langle x_p(\omega)x_{p'}(\omega')\rangle$. In order to calculate the dynamic structure factor following Eq. (32), the time-dependent correlators $\langle x_p(t)x_p(0)\rangle$ are needed.

For small and large ω, explicit expressions may be evaluated analytically. The short-time behavior of the correlators becomes

$$\langle x_p(t)x_p(0)\rangle = \frac{k_B T}{2N\tilde{v}_p}\left(\frac{\Delta\zeta(0)}{\zeta\tilde{v}_p} + e^{-\tilde{v}_p t}\left(1 - \frac{\Delta\zeta(0)}{\zeta\tilde{v}_p}\right)\right) \tag{37}$$

with $\tilde{v}_p = v_p + \Delta\zeta(0)/\zeta$

$v_p = 1/\tau_p$. Besides the Rouse model, where the correlations decay exponentially [Eq. (18)], the effect of entanglement friction leads to an exponential decay of only part of $\langle x_p(t)x_p(0)\rangle$. A time-dependent plateau remains the height of which depends via $\tilde{v}_p$ on the mode number and the relative strength of the entanglement friction rate $\Delta\zeta(0)/\zeta$. The modes do not completely decay; large-scale relaxations with low p are suppressed or strongly reduced.

In the long-time limit one finds

$$\langle x_p(t)x_p(0)\rangle = \frac{k_B T}{2N\zeta v_p} \exp\left(-\frac{v_p}{1 + \dfrac{5}{3}\dfrac{\Delta\zeta(0)}{\zeta}\tau_{\text{eff}}}\right) \tag{38}$$

Using $\tau_{\text{eff}} = (6/5)\,(N/N_c)\zeta/\Delta\zeta(0)$, the relaxation time becomes

$$\tilde{v}_p = v_p(1 + 2N/N_c)^{-1} \tag{39}$$

which in the long-chain limit agrees with the reptation prediction $v \approx 1/N^3$ (see next section). For sufficiently long chains the center of mass diffusion displays an interesting crossover from Rouse-like diffusion at short times ($D \approx 1/N$) to reptation diffusion ($D \approx 1/N^2$, see next section) at long times. In order to visualize a typical time-dependent diffusion coefficient, in Fig. 11 we present the

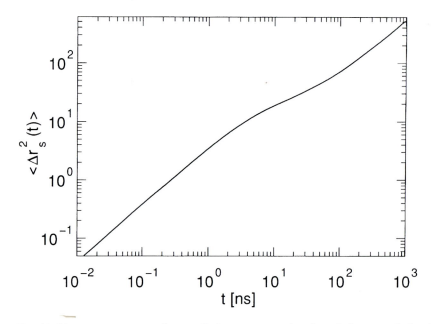

Fig. 11. Mean square center of mass displacement as a function of time as calculated for a $M_W = 6500$ g/mol PE chain on the basis of the parameters obtained from a fit of the spectra to the Hess model. (Reprinted with permission from [36]. Copyright 1994 American Chemical Society, Washington)

predicted mean-square center of mass displacement for a $M_W = 6500$ chain using parameters from an evaluation of experimental spectra to be discussed below. Figure 11 shows that within the experimental time range (between 10 and 100 ns) the crossover from short-to long-time behavior takes place.

3.2.2 NSE Results from the Transition Regime

In order to find out how the onset of entanglement effects influences the intrachain motions, NSE experiments were carried out on a number of poly(ethylene) (PE) melts at $T = 236\,°C$ with molecular masses in the transition range ($M_W = 2000$, 3600, 4800, 6500 g/mol) [36, 52]. The monodisperse, deuterated melts were mixed with 10% protonated chains of the same length. Figure 12 shows experimental spectra obtained on specimens with molecular weights of 2000 and 4800. A comparison of the spectra shows that the relaxation is obviously much less advanced for the longer chain compared to the shorter chain. Let us take the spectrum at the momentum transfer $Q = 0.12\,Å^{-1}$ as an example. While the dynamic structure factor for the shorter chain has already decreased to about 0.1 after 20 ns, the longer chain only relaxes to about 0.4 for the same Q and the same time.

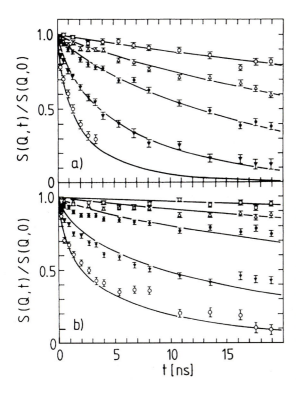

Fig. 12a, b. Dynamic structure factor for two polyethylene melts of different molecular mass; **a** $M_W = 2 \times 10^3$ g/mol **b** $M_W = 4.8 \times 10^3$ g/mol. The momentum transfers Q are 0.037, 0.055, 0.077, 0.115 and 0.155 $Å^{-1}$ from *top* to *bottom*. The *solid lines* show the result of mode analysis (see text). (Reprinted with permission from [36]. Copyright 1994 American Chemical Society, Washington)

Mode Analysis

Following the mode analysis approach described in Section 3.2.1, the spectra at different molecular masses were fitted with Eqs. (32) and (33). Figure 13 demonstrates the contribution of different modes to the dynamic structure factor for the specimen with molecular mass 3600. Based on the parameters obtained in a common fit using Eq. (32), $S(Q, t)$ was calculated according to an increasing number of mode contributions.

Figure 13a shows the contribution of translational diffusion. The translational diffusion only describes the experimental data for the smaller momentum transfer $Q = 0.037 \text{Å}^{-1}$. Figure 13b presents $S(Q, t)$, including the first mode. Obviously, the long-time behavior of the structure factor is now already adequately represented, whereas for shorter times the chain apparently relaxes much faster than calculated.

Figures 13c–e shows how the agreement between experimental data and the calculated structure factor improves if more and more relaxation modes are included. In Fig. 13e, finally, very good agreement between theory and experiment can be noted.

Figure 14 shows the results for W_p, the mode-dependent relaxation rate, for the different molecular masses as a function of the mode index p. For the smallest molecular mass $M_W = 2000 \text{ g/mol}$ relaxation rates W_p are obtained which are independent of p. This chain obviously follows the Rouse law. The modes relax at a rate proportional to p^2 [Eq. (17)]. If the molecular weight is increased, the relaxation rates are successively reduced for the low-index modes in comparison to the Rouse relaxation, whereas the higher modes remain uninfluenced within experimental error.

How can we understand this behavior? In the next section, where it is shown that for polyethylene at the experimental temperature of 509 K the molecular mass between entanglements is $M_e = 2000 \text{ g/mol}$, corresponding to $N_e = M_e/M_0 \approx 140$ monomers (M_0: molar monomer mass). Let us assume that the characteristic length for a relaxation mode L_p is given by the distance between two knots ($L_p = \ell N/p$). We can then define a critical mode index $p_{cr} = N/N_e$ below which the characteristic extension of a mode becomes greater than the distance between entanglements in the long-chain melt. These critical mode indices for the different molecular weights are plotted as arrows in Fig. 14. It is evident that relaxation modes show deviations from Rouse behavior when their extension is greater than the entanglement distance on the long-chain melt. We thus come to an interpretation of the results of mode analysis. Topological interactions block, or at least very much reduce, the relaxation rate for those modes whose characteristic length becomes greater than the entanglement distance formed in long-chain melts.

We are now going to compare the results of mode analysis with measurements of viscosity on polyethylene melts. With the aid of Eq. (30), which links the viscosity to the relaxation times, we can predict the viscosity using the results of spin-echo measurements and compare it with the viscosity measurement

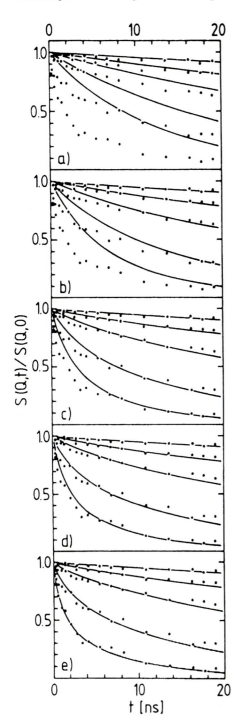

Fig. 13a–e. Result of the mode analysis for the $M_w = 3600$ g/mol sample. The diagrams show the result of a calculation of the spectra retaining a successively rising number of modes in comparison to the experimental result; **a** translation diffusion only, **b** translation diffusion and first mode, **c** translation diffusion and the first two modes, **d** translation diffusion and the first three modes, **e** translation diffusion and all modes. (Reprinted with permission from [36]. Copyright 1994 American Chemical Society, Washington)

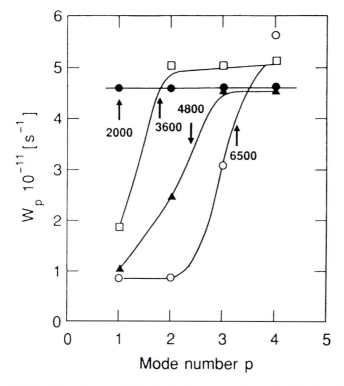

Fig. 14. Relaxation rates W_p for the first four relaxation modes of chains of different molecular mass as a function of the mode number p. The *arrows* indicate the condition $p = N/N_e$ for each molecular weight. (Reprinted with permission from [52]. Copyright 1993 The American Physical Society, Maryland)

[36,47]. This is done in Fig. 15 where the viscosity is shown as a function of molecular mass. The open circles represent the predictions of the NSE result, whereas the filled circles represent the viscosity measurement. Both data sets are in excellent agreement and demonstrate the consistency of evaluation.

Generalized Rouse Model

In addition to the Rouse model, the Hess theory contains two further parameters: the critical monomer number N_c and the relative strength of the entanglement friction $\Delta\zeta(0)/\zeta$. Furthermore, the change in the monomeric friction coefficient with molecular mass has to be taken into account. Using results for $\zeta(M)$ from viscosity data [47], Fig. 16 displays the results of the data fitting, varying only the two model parameters N_c and $\Delta\zeta(0)/\zeta$ for the samples with the molecular masses $M_W = 3600$ and $M_W = 6500$ g/mol.

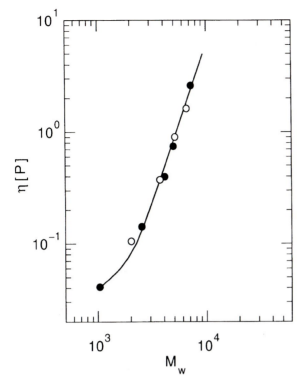

Fig. 15. Comparison of the viscosities either directly measured or calculated from the spin-echo results for polyethylene melts at 509 K as a function of molecular mass (● experimental result; ○ viscosities calculated on the basis of mode analysis). (Reprinted with permission from [52]. Copyright 1993 The American Physical Society, Maryland)

In this fitting procedure the time-dependent diffusion coefficient as discussed above has been taken explicitly into account. As can be seen, with only two parameters both sets of spectra are well described. This holds also for the other two samples. For the model parameters, the fit reveals $N_c = 150$ and $\Delta\zeta(0)/\zeta = 0.18$ ns. N_c is very close to $N_c = 138$ as obtained from the NSE experiments on long-chain PE (see next section).

According to Hess, the relative strength of the entanglement friction can be related to the more microscopic parameter q_c^{-1}, describing the range of the true interchain interaction potential. A value of $q_c^{-1} = 7\,\text{Å}$, close to the average interchain distance of about $4.7\,\text{Å}$, is obtained.

Thus, with only two parameters, the values of which are both close to expectations, the Hess model allows a complete description of all experimental spectra. In the complex crossover regime from Rouse motion to entanglement controlled behavior, this very good agreement confirms the significant success of this theory.

Figure 17 displays the time-dependent correlations $\langle x_p(t)x_p(0)\rangle$ calculated for the $M_w = 6500\,\text{g/mol}$ sample on the basis of the fit results. The curves present the time-dependent relaxation of the different modes, commencing with $p = 1$ at the top. In order to separate the different curves, they have been plotted

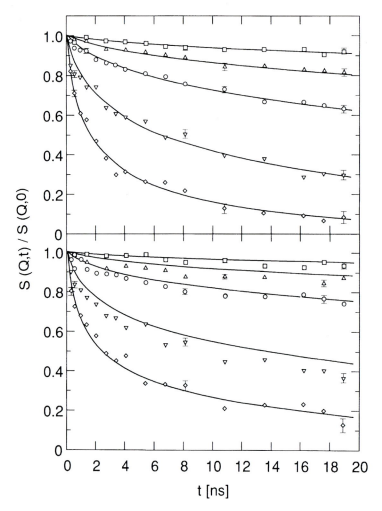

Fig. 16. Experimental spectra from the PE sample with $M_w = 3600$ g/mol (*above*) and $M_w = 6500$ g/mol (*below*) compared to the calculated spectra from the Hess model where the parameters were obtained from a joint fit of all spectra. (Reprinted with permission from [36]. Copyright 1994 American Chemical Society, Washington)

to include the $1/p^2$ prefactor [Eq. (32)]. While the first modes relax only marginally, the main relaxation of the dynamic structure factor on the experimental time window is observed on the higher p modes. The two-step relaxation exhibited by the higher p modes reflects the relaxation behavior displayed by Eq. (37). The correlation functions first decay to a plateau, the value of which depends on the mode number. For long times final relaxation occurs.

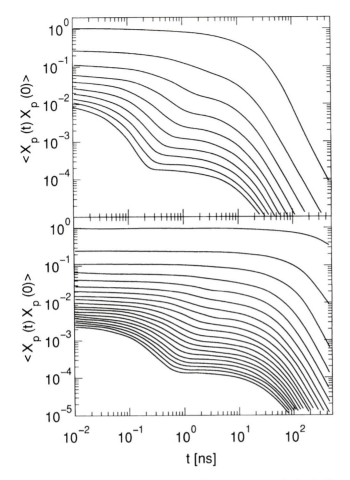

Fig. 17. Time-dependent correlators for different Rouse modes in the Hess model. The calculations where performed for PE of $M_w = 6500$ g/mol. The mode numbers increase from the *top* commencing with p = 1. (Reprinted with permission from [36]. Copyright 1994 American Chemical Society, Washington)

In conclusion, the Q-dependence of the mode contributions to the dynamic structure factor facilitates direct access to the individual relaxation modes of a chain. This follows from the fact that neutrons permit motion to be observed on their natural length and time scales. We find that large-scale modes with a characteristic length greater that the entanglement distance or with a mode index $p < N/N_c$ are slowed down significantly. Quantitative agreement with viscosity measurements is achieved with the aid of the extracted relaxation rates. A comparison with the generalized Rouse model according to Hess demonstrated that this model is able to describe the Q and molecular weight dependence of the dynamic structure factor very well.

3.3 Entanglement Constraints in Long-Chain Polymer Melts

3.3.1 Theoretical Outline – The Reptation or Tube Concept

As mentioned at the beginning of Section 3, the plateau modulus of long- chain polymer melts has been interpreted in analogy to rubber elasticity by the hypothesis of a temporary network built from the entanglements of the mutually interpenetrating chains. A number of theories of viscoelasticity have been developed on the basis of this assumption. The most famous of these is the reptation model by de Gennes and Doi and Edwards [8, 10]. In this model, a snake-like creep along the chain profile is postulated as the dominant chain movement. The lateral freedom of this is modeled by a tube confinement following the coarse-grained chain profile. The tube diameter d is identified by the spacing of the cross-links of the rubber analogue. This establishes a relation between the plateau modulus and the tube diameter [5]

$$G_N^0 = \frac{4}{5} \frac{\langle R_e^2 \rangle}{M} \frac{\rho kT}{d^2} \tag{40}$$

where $\langle R_e^2 \rangle = nNC_\infty \ell_0^2$ is the mean square end-to-end distance, C_∞ is the characteristic ratio, ℓ_0 the average main chain bond length, n the average bond number/monomer, and N the number of monomers in the chain. Equation (40) represents one of the fundamental relationships of the Doi-Edwards theory relating macroscopic viscoelastic properties to microscopic chain confinement.

Tube confinement leads to strong alterations of the mean square segment displacements as compared to the Rouse model.

1. *Rouse regime*: For short times, during which a chain segment does not regard any of the topological restrictions of the movement (r < d), we expect *unrestricted Rouse behavior* $\Delta r_{nn}^2(t) \sim t^{1/2}$ [Eq. (20)]. This motional behavior reaches its limits if Δr_{nn}^2 becomes comparable with the tube diameter d^2. With the aid of Eq. (20) and the condition $\Delta r_{nn}^2 = d^2$, it follows for the crossover time

$$\tau_e = \frac{\pi}{12} \frac{d^4 \zeta}{k_B T \ell^2} \tag{41a}$$

A different estimate, which identifies τ_e with the slowest Rouse mode of a chain with the end-to-end distance, $R_e^2 = d^2$ leads to a quite similar result

$$\tau_e = \frac{\pi}{3\pi^2} \frac{d^4 \zeta}{k_B T \ell^2} \tag{41b}$$

2. *Local reptation regime*: For times $t > \tau_e$ we have to consider curvilinear Rouse motion along the spatially fixed tube. The segment displacement described by Eq. (18) (n = m) must now take the curvilinear coordinates s along the tube into consideration. We have to distinguish two different time regimes. For $(t < \tau_R)$, the second part of Eq. (19) dominates – when the Rouse modes

equilibrate along the tube, we retain the familiar $t^{1/2}$ law – while for $t > \tau_R$, the first term carries the weight – the chain diffuses along the tube. If a segment performs a mean square displacement $\langle (s(t) - s(0))^2 \rangle$ along the tube where s are curvilinear coordinates, its mean square displacement in three-dimensional space is $\langle (r(t) - r(0))^2 \rangle \cong d \langle (s(t) - s(0))^2 \rangle^{1/2}$. The tube itself constitutes a Gaussian random walk with step length d. For the two regimes we thus obtain [34]

$$\Delta r_{nn}^2(t) = d \left(\frac{k_B T \ell^2}{\zeta} t \right)^{1/4} \qquad \tau_e < t < \tau_R$$

$$\Delta r_{nn}^2(t) = d \left(\frac{k_B T \ell^2}{N \zeta} t \right)^{1/2} \qquad \tau_R < t < \tau_d \tag{42}$$

where τ_d is the terminal time at which the chain has completely left its original tube.

3. *Diffusion regime*: For times $t > \tau_d \cong N^3 \zeta l^4 / k_B T d^2$, the dynamics are determined by reptation diffusion. We expect normal diffusive behavior

$$\Delta r_{nn}^2(t) \cong \frac{k_B T d^2}{N^2 \ell^2 \zeta} t \tag{43}$$

The reptation model thus predicts four dynamic regimes for segment diffusion. They are summarized in Fig. 18.

Dynamic Structure Factors

As mentioned in Section 3.1, the incoherent dynamic structure is easily calculated by inserting the expression for the mean square displacements [Eqs. (42), (43)] into Eq. (4b). On the other hand, for reptational motion, calculation of the pair-correlation function is rather difficult. We must bear in mind the problem on the basis of Fig. 19, presenting a diagrammatic representation of the reptation process during various characteristic time intervals.

1. *Rouse regime*: For $t < \tau_e$, we are dealing with unrestricted Rouse movements. Here, the dynamic structure factor of the Rouse model should be valid. For $t = \tau_e$, density fluctuations are equilibrated across the tube profile, and we enter the regime of (2) local reptation [53]. Density fluctuations equilibrate within the chain along the fixed tube. Here, the $t^{1/4}$ law is valid for the self-correlation function. The pair-correlation function is sensitive to changes in the relative position of different segments. During Rouse relaxation in the tube, the relative arrangements of segments changes only marginally, they slip to a large extent collectively up and down the tube. For the dynamic structure factor this means that S(Q, t) tends to approach a Q-dependent plateau value. For $\tau_R < t < \tau_d$, the chain creeps out of its tube. Here, τ_d is the time after which the chain has completely left its original tube. Correlations between the various

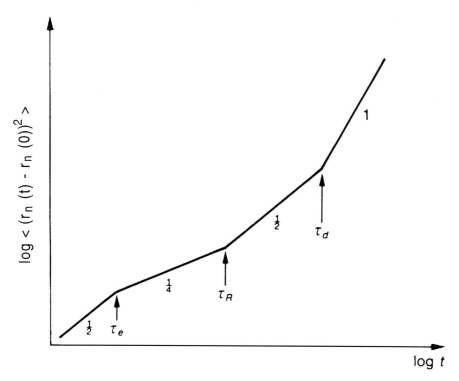

Fig. 18. Mean square displacement of a chain segment undergoing reptational motion as a function of time

chain domains are gradually lost. We expect a further decrease in $S(Q, t)$. For $t > \tau_d$, the segment diffusion process is converted into reptation diffusion. Normal diffusive behavior, determined by the reptation diffusion constant, can be expected here.

2. *Regime of local reptation*: We have already dealt with domain (1). For the regime of local reptation ($t > \tau_e$) a calculation by de Gennes gives $S(Q, t)$ to the first order of $(Qd)^2$ and neglects the initial Rouse movement [53]:

$$S(Q, t) = 1 - \frac{Q^2 d^2}{36} + \frac{Q^2 d^2}{36} \exp\left(\frac{u^2}{36}\right) \mathrm{erfc}\left(\frac{u}{6}\right) \tag{44}$$

where u is the 'Rouse variable' (see Table 2). As discussed above, for long times, $S(Q, t)$ approaches a plateau: $S(Q, t \rightarrow \infty) = 1 - Q^2 d^2/36$. The existence of the tube diameter as a dynamic characteristic length invalidates the scaling property of the Rouse model, $S(Q, t) = S((Q\ell)^2 (Wt)^{1/2})$ and leads to characteristic deviations. Apart from the dependence on $(Q\ell)^2 (Wt)^{1/2}$, $S(Q, t)$ depends explicitly on Qd. This result is of a general nature and does not depend on the special de Gennes model.

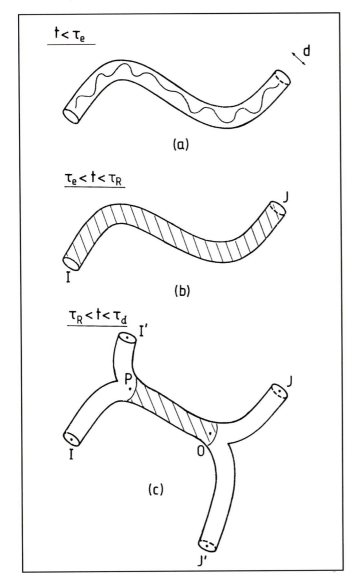

Fig. 19a–c. Schematic representation of a reptating chain in different time regimes **a** Short-time unrestricted Rouse motion; **b** equilibration of density fluctuations along tha chain; **c** creep motion of a chain out of its tube.

A quantitative analysis of scattering data, originating from the crossover regime between short-time Rouse motion and local reptation, needs explicit consideration of the initial Rouse motion neglected by de Gennes. Ronca [50] proposed an effective medium approach, where he describes the time-dependent

Table 2. Dynamic structure factors for reptational dynamics

	Coherent dynamic structure factor	Adjustable parameters
Long-time behaviour of local reptation (de Gennes [52])	$$S(Q,t)/S(Q,0) = 1 - \frac{Q^2d^2}{36} + \frac{Q^2d^2}{36}\exp\frac{u^2}{36}\,\text{erfc}\,\frac{u}{6}$$ $u = Q^2\ell^2(Wt)^{1/2} = 6(\Omega_R t)^{1/2}$: input parameter	d
Local reptation with inclusion of Rouse relaxation (Ronca [49])	$$S(Q,t)/S(Q,0) = \frac{Q^2d^2}{24}\int_0^\infty dx\,\exp\left\{-\left(\frac{Q^2d^2}{48}\,g\left(x,\left(\frac{16W\ell^4 t}{d^4}\right)\right)\right)\right\}$$ $g(x,y) = 2x - \exp(x)\,\text{erfc}\,\{y^{1/2}+x/(4y)^{1/2}\} + \exp(-x)\,\text{erfc}\,\{x/(4y)^{1/2} - y^{1/2}\}$	$\Omega_R(Q)$, d $\Omega_R(Q) = \dfrac{Q^4\ell^4}{36}W,$
Long-chain dynamics in the presence of fixed stress points (entanglements) (des Cloizeaux [55])	$$S(Q,t)/S(Q,0) = \frac{1}{Z}\ln(1+Z) + \int_0^\infty dp\,e^{-p}\,F(X,pZ)$$ $Z = q^2d^2/6 \qquad X = \pi^{1/2}u/6$ $$F(X,pZ) = pZ\int_0^1 dz\int_0^1 dv\,\exp-\left\{pZz + \pi^{-1/2}\int_0^x dw\,B(z,v,w,pZ)\right\}$$ $$B(z,v,w,pZ) = \sum_{p=-\infty}^{\infty}\exp-\left\{\frac{(z-2p)^2Z^2}{w^2}\right\} - \exp-\left\{\frac{(v-2p)^2Z^2}{w^2}\right\}$$	$\Omega_R(Q)$, d

friction in terms of a phenomenological memory function. His approach has similarities to the more microscopic treatment of entanglement friction discussed above. In the particular case of infinite chains, both treatments result in the same expression for the dynamic structure factor. In order to model the chain confinement, Ronca introduced an additional viscoelasticity term in the Langevin equation for the Rouse movement:

$$\zeta \frac{\partial r_n}{\partial t} = \frac{3 k_B T}{\ell^2} \frac{\partial^2 r_n}{\partial n^2} + d \int_\infty^t K(t - t') \frac{\partial r_n}{\partial t'} dt' + f_n(t) \tag{45}$$

The memory function $K(t)$ phenomenologically describes the elastic coupling between the chains. Equation (45) is similar to viscoelastic equations involving the time-dependent modulus $G(t)$. Phenomenologically, it was argued that $K(t)$ must therefore be related to $G(t)$. If each segment was to move like a macroscopic object in the medium of the other chains, then $K(t) \equiv G(t)$ would be valid. For infinite chains, the solutions of Eq. (45) have the form of Eq. (37). In particular, for the transition regime from Rouse motion to entanglement controlled dynamics, the mean square displacements between different segments becomes

$$\langle (r_n(t) - r_m(0))^2 \rangle = \frac{d^2}{24} g \left(4 \frac{|n - m|}{N_e}, \frac{16 W t}{N_e^2} \right) \tag{46}$$

with

$$g(x, y) = 2x - \exp(x) \operatorname{erfc} \left(\sqrt{y} + \frac{x}{\sqrt{2y}} \right) + \exp(-x) \operatorname{erfc} \left(\frac{x}{\sqrt{2y}} - \sqrt{y} \right)$$

where $\tau_e = \zeta d^4 / 48 k_B T \ell^2$ is the characteristic time constant for the entanglement transition. Within a factor of 2 this transition time agrees with expression (41b) defining τ_e as the Rouse time of an entanglement strand. Inserting Eq. (46) into Eq. (4a) yields the dynamic structure factor $S(Q, t)$ for the Ronca model (see Table 2):

$$\frac{S(Q, t)}{S(Q)} = \frac{Q^2 d^2}{24} \int_0^\infty \exp \left[-\frac{Q^2 d^2}{48} g \left(z, \frac{16 W \ell^4 t}{d^4} \right) \right] dz \tag{47}$$

Like the dynamic structure factor for local reptation it develops a plateau region, the height of which depends on Qd. Figure 20 displays $S(Q, t)$ as a function of the Rouse variable $Q^2 \ell^2 \sqrt{Wt}$ for different values of Qd. Clear deviations from the dynamic structure factor of the Rouse model can be seen even for $Qd = 7$. This aspect agrees well with computer simulations by Kremer et al. [54, 55] who found such deviations in the Q-regime $2.9 \leqslant Qd \leqslant 6.7$.

Recently, des Cloizeaux has conceived a 'rubber'-like model for the transition regime to local reptation [56]. He considered infinite chains with spatially fixed entanglement points at intermediate times. In between these

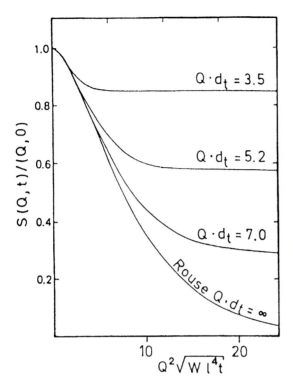

Fig. 20. Single-chain dynamic structure factor of the Ronca model as a function of the Rouse variable for different values of Qd_t (d_t tube diameter; $d_t \equiv d$). (Reprinted with permission from [50]. Copyright 1983 American Institute of Physics, Woodbury N.Y.)

temporary 'cross-links' the chains are allowed to perform Rouse motion. For the distribution of entanglement points along a given chain, a Poisson distribution was assumed. The explicit form of the normalized, coherent dynamic structure factor is given in Table 2. Qualitatively, the des Cloizeaux and the Ronca models arrive at similar results for $S(Q, t)$ in the transition region. Quantitatively, as a consequence of the assumed distribution of entanglement distances, the Q-dispersion of the plateau heights is less pronounced as in the Ronca model. Furthermore, for a given averaged entanglement distance $S(Q, t)$ in the rubber-like model decays less than for the effective medium or the local reptation model. This may be rationalized considering that there the entanglement points are fixed in space.

3. *Regime of creep*: In the time range $t > \tau_R$, the chain creeps out of the tube. Doi and Edwards give a simple argument for the shape of the dynamic structure factor for the range $QR_G \gg 1$ [5]. For the parts of chain still in the original tube $S(Q, t) = 12/Q^2\ell^2$ and for those parts of the chain which have already crept out of the tube all correlations have subsided and $S(Q, t) = 0$. It follows

$$S(Q, t) = \frac{12}{Q^2\ell^2} \Psi(t) \tag{48}$$

where

$$\Psi(t) = \frac{8}{\pi^2} \sum_{p_{odd}} \frac{1}{p^2} \exp\left(-\frac{tp^2}{\tau_d}\right)$$

is the probability that a piece of the chain will remain in the tube.

4. *Diffusion regime*: Finally, for time $t > \tau_d$ and $QR_G < 1$, translational diffusion of the whole chain is observed resulting in

$$S(Q,t) = Ne^{-Q^2Dt} \tag{49}$$

with the diffusion coefficient for reptation $D = 1/3 \, k_B T d^2/N^2\ell^2\zeta$.

The presence of entanglement constraints is expected to show itself (1) by a reduction in the time decay of $S(Q,t)$ compared to the Rouse dynamic structure factor and (2) by the systematic Q-dependent deviation from the Rouse scaling properties.

3.3.2 NSE Studies on Entangled Polymer Melts

Poly(tetrahydrofurane) (PTHF)

First evidence for a slowing down of $S(Q,t)$ decay was reported by Higgins and Roots [57], who compared dynamic structure factors obtained from 'long' labelled PTHF chains in matrices of short and long chains. Figure 21 compares $S(Q,t)$ at $Q = 0.09 \, \text{Å}^{-1}$ for both samples. Clearly, in the long-chain environment, $S(Q,t)$ decays much more slowly than among short chains. Unfortunately, in this experiment the polymers studied were ill defined with M_W-distributions ranging from 10^2 to 2.5×10^4 g/mol for the short chains and 10^3 to 10^6 g/mol for

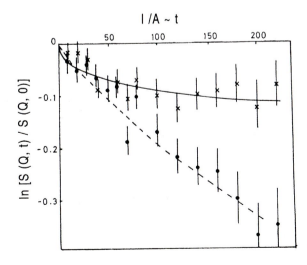

Fig. 21. Comparison of the dynamic structure factors from long PTHF chains in a matrix of long chains ($\times$) with that in a matrix of short chains ($\bullet$). The Q-value of the experiment was $Q = 0.09 \, \text{Å}^{-1}$, the temperature $T = 418$ K (Reprinted with permission from [57]. Copyright 1985 Royal Society of Chemistry, Cambridge, UK)

the long chains. Under these circumstances the friction coefficients of the matrices will be quite different [47], and part of the effect must be traced to changes in the general mobility. Consequently, model-independent signatures for the dynamic length like systematic deviations from Rouse scaling could not be established. Nevertheless, Fig. 21 shows that a long-chain environment clearly restricts the chain mobility in comparison to the short-chain mixture.

Poly(isoprene) (PI)

The first experiments clearly showing systematic Q-dependent deviations from Rouse scaling were performed on PI [39]. They provided first evidence for the existence of the intermediate dynamic length scale set by the entanglement distance d. As already displayed in Fig. 6 for a PI melt at 200 °C ($M_W = 52\,000$ g/mol, $M_W/M_n = 1.02$) in a time window of 20 ns, the decay of $S(Q, t)$ is systematically weaker compared to the Rouse dynamic structure factor. Figure 22 displays the same spectra in a scaling representation using $W\ell^4 = 2.0 \cdot 10^{13}\,\mathring{A}\,s^{-1}$, the value found from the short-time Rouse regime. For

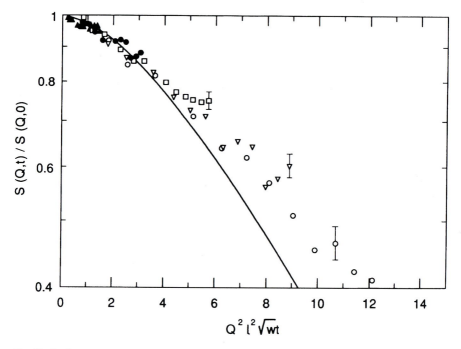

Fig. 22. Scaling representation of the spectra obtained from a Polyisoprene melts in terms of the Rouse variable. The *solid line* displays the Rouse structure factor using the Rouse rate determined at short times. ($\circ$ $Q = 0.153\,\mathring{A}^{-1}$; $\triangledown$ $Q = 0.128\,\mathring{A}^{-1}$; $\square$ $Q = 0.102\,\mathring{A}^{-1}$. $\bullet$ $Q = 0.077\,\mathring{A}^{-1}$; $\triangle$ $Q = 0.064$ $\mathring{A}^{-1}$) (Reprinted with permission from [39]. Copyright 1992 American Chemical Society, Washington)

small values of the scaling variable u, the data are reasonably well described by the Rouse master function (look, however, for the long-time results at $Q = 0.064$ Å^{-1}). Towards larger u the data fall systematically above the master function indicating slower relaxation or the presence of constraints. Within experimental accuracy, however, the deviations from scaling are still marginal though slight indications for a splitting of the spectra with Q are suggestive. In this respect, the spectra are similar to those obtained from PDMS, where at larger values of the scaling variable deviations slower than Rouse relaxation were seen [41, 42].

Figure 23 presents spectra from a PI sample with somewhat larger molecular mass ($M_W = 79\,100$ g/mol; $M_W/M_n = 1.03$) again taken at 200 °C. For this experiment the time range has been extended to $t_{max} = 46$ ns. Displayed in a Rouse scaling plot these data exhibit a pronounced and systematic splitting with Q. The solid lines represent a fit with Ronca's effective medium model [50]. Again, the Rouse rate was fixed to the value obtained at short times. The only parameter fitted was the entanglement distance yielding $d = 52 \pm 1$ Å (see also Table 3). Overall, Ronca's model provides a good description of the experimental data concerning both the line shape as well as the magnitude of the

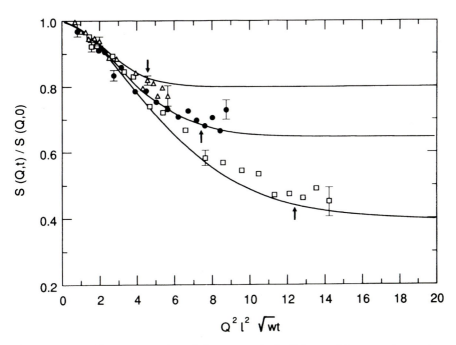

Fig. 23. Rouse-scaling representation of the spectra obtained from polyisoprene at a neutron wavelength of $\lambda = 11.8$ Å and T = 473 K ($\triangle$ Q = 0.074 Å^{-1}; $\bullet$ Q = 0.093 Å^{-1}; $\square$ Q = 0.121 Å^{-1}). The *solid lines* are the result of a fit with the Ronca model [50]. The *arrows* indicate $Q^2\ell^2\sqrt{W\tau_e}$ for each curve. (Reprinted with permission from [39]. Copyright 1992 American Chemical Society, Washington)

Table 3. NSE results on polymer melts

			NSE		Rheology
Sample	T(K)	d(Å)		$\tau_e(ns)^a$	d(Å)
PI	473	52 ± 1^b		32	51 (298 K)
PEP	492	47.5 ± 0.4^b		15	43.1 ± 2
homopolymer		49.6 ± 0.9^c			
PEP triblock	491	47.1 ± 0.7^b		15	
PEB-2	509	43.5 ± 0.7^c			35 (373 K) PE
		43.1 ± 0.9^b		5	42 (509 K) PEB-7
		60.6^d			
PDMS	473	70 ± 3^b		60	75 (413 K)

[a] Calculated on the basis of Eq. (41b).
[b] Fit with the Ronca model.
[c] Fit with local reptation.
[d] Fit with the des Cloizeaux model.

Q-splitting. Using Eqs. (41a) and (41b) the crossover time τ_e from unrestricted Rouse to entanglement controlled behavior was calculated. The calculation on the basis $\Delta r_{nn}^2 = d^2$ [Eq. (41a)] yields $\tau_e = 250$ ns, clearly overestimating τ_e. In this case, entanglement effects would have hardly been seen in the experimental time window ($t_{max} = 46$ ns). Equation (41b) which calculates τ_e as the Rouse time of a chain between entanglements gives $\tau_e = 32$ ns which is compatible with the experimental spectra. Small arrows in Fig. 23 indicate the position in the scaling plot where $Q^2 \ell^2 \sqrt{W\tau_e} = (Qd)^2/\pi$ is reached, showing that experimental time range just reaches the crossover region. Finally, in terms of Ronca's model the characteristic crossover time is $\tau_e = 20$ ns.

Alternating (Polyethylene-Propylene) (PEP) Copolymer

In the case of PEP, a homo- and a triblock copolymer ($M_W = 83\,800$ g/mol; $M_W/M_n = 1.03$; triblock: $M_W = 87\,800$ g/mol; $M_W/M_n = 1.03$), where in the latter case the middle block was labelled, have been investigated [39, 58, 59]. Figure 24 displays the measured dynamic structure factors for the two polymers at T = 492 K. The scattering functions are characterized by a rapid initial decrease followed by longer times by a plateau regime in time. The observation of an "elastic" time-independent contribution to S(Q,t) is the main result of this experiment. For comparison for the largest Q the dashed line in Fig. 24 displays the dynamic structure factor for unrestricted Rouse relaxation exhibiting the same initial decay. A comparison of the homopolymer and triblock results shows that, apart from a slightly slower relaxation in the initial decay, the triblock has a similar relaxation behavior. In particular, the long-time plateaus are identical for both polymers, indicating that they both relax under the same constraints, whereby end effects appear to be of no importance.

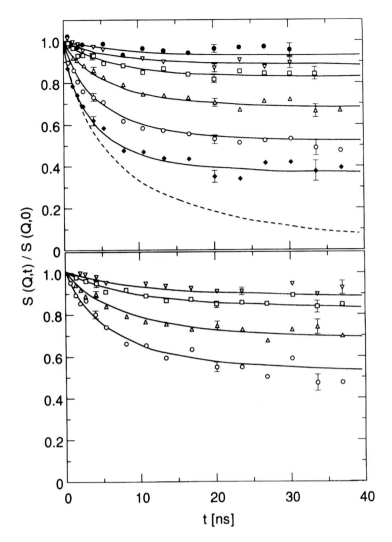

Fig. 24. NSE spectra from the PEP homopolymer (*above*) and the triblock (*below*) at 492 K. The *solid lines* are the result of a fit with the Ronca model [49]; the *dashed line* presents the expected dynamic structure factor for Rouse relaxation corresponding to the highest Q-value. (● Q = 0.058 Å^{-1}; ▽ Q = 0.068 Å^{-1}; □ Q = 0.078 Å^{-1}; △ Q = 0.097 Å^{-1}; ○ Q = 0.116 Å^{-1}; ◆ Q = 0.116 Å^{-1}). The *arrows* mark the crossover time τ_e. (Reprinted with permission from [39]. Copyright 1992 American Chemical Society, Washington)

Figure 25 presents the homopolymer data in a scaled form. The splitting into Q-dependent plateau levels is much more pronounced than for PI, a phenomenon resulting from a faster relaxation rate and somewhat stronger constraints. Again, the Ronca model represents an excellent description of the experimental data reproducing the line shape, i.e. the relatively sharp crossover and the

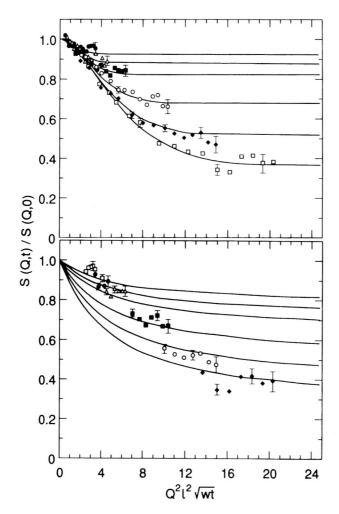

Fig. 25. Rouse-scaling representation of the PEP homopolymer data at 492 K. *Above solid lines* represents the Ronca model [50]; *below solid lines* display the predictions of local reptation [53]. The *solid lines* from *below to above* correspond to Q = 0.135 Å^{-1}; Q = 0.116 Å^{-1}; Q = 0.097 Å^{-1}; Q = 0.078 Å^{-}; Q = 0.068 Å^{-1}; Q = 0.058 Å^{-1}). The *symbols* along the lines are data points corresponding to the respective Q-value. (Reprinted with permission from [39]. Copyright 1992 American Chemical Society, Washington)

Q-dependence of the plateau levels. The results for the model parameters are given in Table 3. The crossover time τ_e = 15 ns (T = 492 K) calculated with Eq. (41b) agrees well with the observed spectral shape in dividing the initial fast decay from the plateau-like behavior at longer times. Figure 25 (lower part) compares the experimental spectra for times longer than τ_e with the local reptation model of de Gennes [Eq. (44)]. This model omits the initial Rouse decay, therefore, data for $t < \tau_e$ were not considered. Furthermore, the Rouse rate, which is determined by the initial decay of the spectra, was imposed on the fit. As can be seen, the long-time behavior of the experimental results is well depicted, making it impossible to distinguish between the two models.

Polyethylene (PE)

With respect to the intensity resolution relationship of NSE, PEB-2 [essentially PE with one ethyl branch every 50 main chain bonds; the sample is obtained by saturating 1–4 polybutadiene, the residual 1–2 groups (7%) cause the ethyl branches; $M_w = 73200$ g/mol; $M_w/M_n = 1.02$] has two advantages compared to PEP: (1) the Rouse rate $W\ell^4$ of PEB-2 is more than two times faster than that of PEP at a given temperature [$W\ell^4_{PEP}$ (500 K) $= 3.3 \times 10^{13}$ Å^4 s^{-1}; $W\ell^4_{PEB}$ (509 K) $= 7 \times 10^{13}$ Å^4 s^{-1}]; (2) at the same time, the topological constraints are stronger.

   Therefore, it was possible to access the interesting time regime using neutrons of shorter wavelength ($\lambda = 8.5$ Å) covering the time window $0.3 \leqslant t \leqslant 17$ ns [60]. The neutron intensity at this wavelength is about four times higher than that at $\lambda = 11.25$ Å used for PEP, leading to data of higher statistical accuracy. Figure 26 presents representative spectra obtained at 509 K in a scaling form. The solid line in the upper part of Fig. 26 is the result of a fit with the Ronca model. As is evident, the agreement between the Ronca model and the experimental results is not as good as for the other two polymers; the spectra appear to exhibit some systematic deviations from the theoretical curves. For $Q = 0.0078$ Å^{-1} and $Q = 0.116$ Å^{-1} at long times the data points fall below the theoretical line, while for $Q = 0.155$ Å^{-1} the opposite trend is evident. The lower part of Fig. 26 displays a comparison with the local reptation model again restricted to times $t > \tau_e = 5$ ns. As in the case of PEP, the Rouse rate was taken from the initial decay of $S(Q,t)$. Apparently, the local reptation model fits the spectra significantly better than the Ronca model. In particular, it accounts for the persisting gradual decay of $S(Q,t)/S(Q,0)$ even at large $Q^2\ell^2\sqrt{Wt}$. Furthermore, it also describes the amount of Q splitting more accurately than the Ronca model. The resulting entanglement distance d from the fit with the local reptation model differs only slightly from the d-value obtained from the Ronca model (see Table 3).

   Finally, Fig. 27 compares the PE-data with the predictions of the rubber-like model of des Cloizeaux [51]. This model describes the experimental data very appropriately over the entire range of the scaling variable observed. In particular, we note that compared to the Ronca model description the predicted Q splitting is less pronounced and much closer to the experimental observation. In his analysis des Cloizeaux derives a mean distance squared between entanglement of $\langle S^2 \rangle = 12.55$ nm^2. In order to compare results with the entanglement

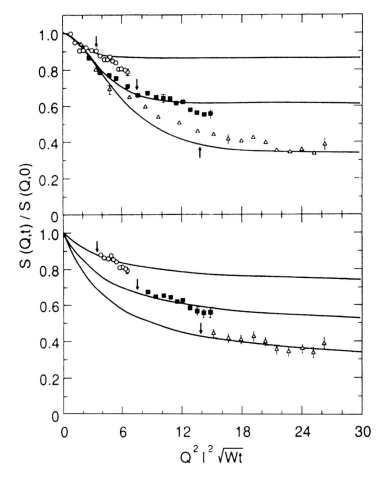

Fig. 26. NSE spectra in polyethylene melts at 509 K in a Rouse scaling plot: ($\circ$ Q = 0.078 Å^{-1}; ■ Q = 0.116 Å^{-1}; △ Q = 0.155 Å^{-1}. *Above* Spectra in comparison to a fit of the generalized Rouse model [49]. *Below* comparison of the data with the predictions of the local reptation model [53] omitting the measurement points which correspond to the initial Rouse relaxation. The arrows indicate $Q^2 \ell^2 \sqrt{W\tau_e}$. (Reprinted with permission from [39]. Copyright 1992 American Chemical Society, Washington)

distances and tube diameters as defined here, this value has to be multiplied by 3, resulting in d = 60.6 Å, a value considerably larger than d = 43 Å as obtained from the other two models. As discussed above, the assumption of spatially fixed entanglement points requires a larger average entanglement distance for a given decay of S(Q, t).

So far, the existence of a well-defined entanglement length in dense polymer systems has been inferred indirectly from macroscopic experiments like measurements of the plateau modulus. However, its direct microscopic observation remained impossible. The difficulty in directly evaluating the entanglement

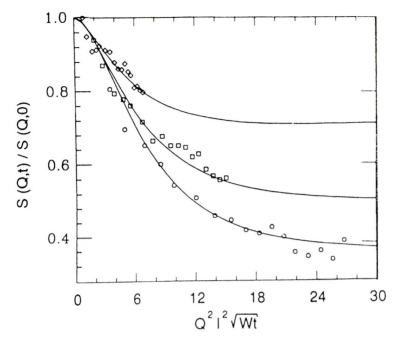

Fig. 27. Comparison of the polyethylene data with the rubber-like model of des Cloizeaux [56]. The Q-values correspond to those in Fig. 26

length is based on its very nature as a dynamic length scale which cannot be traced in static, structural investigations like small-angle scattering. Its observation needed dynamic experiments covering the relevant length and time scales. As discussed in Section 3.3.1, an intermediate dynamic length scale is expected to cause systematic Q-dependent deviations from Rouse scaling because besides the combination $\ell^2 Q^2 \sqrt{Wt}$, it introduces a second explicit Q-dependence into the dynamic structure factor $S(Qd, Q^2 \ell^2 \sqrt{Wt})$. Figures 23 (PI), 25 (PEP), and 26 (PE) display the 'Q-splitting' of $S(Q, t)$ for the three polymers discussed here and prove directly the microscopic existence of an intermediate dynamic length scale which profoundly affects the relaxation of density fluctuations.

Table 3 summarizes the values for the entanglement lengths taken from NSE experiments on those polymers where from a Q-splitting in a Rouse scaling presentation an intermediate dynamic length scale could be directly inferred. Following the reptation model, they are compared with entanglement distances derived from rheological measurements according to Eq. (40). For all polymers the agreement between microscopic and macroscopic results is good. While for PI, PEP and PDMS the agreement is nearly quantitative, the slightly larger difference for PEB-2 compared to PE may be related to the residual ethyl branches (2 of 100 main chain bonds) which cause some extra bulkiness.

A comparison with the tube diameter derived from plateau moduli measurements on PEB-7 underlines this assertion. The coincidence of tube diameters determined macroscopically by application of the reptation model and direct microscopic results is far better than what could have been expected and strongly underlines the basic validity of the reptation approach.

In conclusion, the segment dynamics of long entangled chains in the melt of several polymers by means of NSE spectroscopy was investigated. These measurements led to the direct observation of the mean entanglement distance or tube diameter in long-chain melts. This quantity is thus not only a rheological conceptual model, but indeed exists on the molecular scale. The values of the microscopically determined tube diameters are in excellent agreement with the entanglement distances derived from a reptation interpretation of the rheologically determined plateau moduli. We note that this by no means trivial result interlinks measurements which, on the one hand, were performed on the ns scale (NSE) and, on the other hand, in the kHz range (rheology).

3.4 On the Origin of Entanglement Formation

3.4.1 Models for Entanglement Formation

Although the entanglement distance is of fundamental significance for the understanding of the viscoelasticity of polymer melts, its molecular origin is little understood. Current models interpret it above all as a topological phenomenon orginating from the fact that long, linear, coiled polymer chains cannot intersect one another [61–73]. In such an interpretation the chain contour length density is a decisive parameter, whereby contour length density is understood to be the total chain length to be accommodated in a unit volume. Slender chains without side groups, such as polyethylene, have a high contour length density, whereas e.g. chains with voluminous side groups such as polystyrene are characterized by a low contour length density.

The Scaling Approach. The dominance of the contour length density was the basis of a scaling approach by Graessley and Edwards [61], which we will discuss in the following. De Gennes [65] and others [64] suggested that the number of binary interchain contacts was essential for the occurrence of entanglements. Ronca [50] and others [62, 68] formulated a packing criterion to explain the entanglements. We will see that both concepts can be understood as special cases of the scaling model. The starting point of the scaling model is the assumption that the long-range interaction in dense polymer systems is only related to the linear structure of chains unable to intersect one another. The two parameters of the model are the contour length density (chain length per volume L/V) and the step length of the random walk of the chain. This so-called Kuhn length

$$\ell_K = C_\infty \ell_0 \tag{50}$$

describes the local stiffness of the polymer; the greater the characteristic ratio C_∞, the stiffer is the chain locally. In a polymer network, the modulus reflects the cross-link density. In analogy to the rubber, the plateau modulus G_N^0 is a measure of the interaction density of the chains and should be basically determined by the contour length density. With the contour length density $L/V = vL$ ($v = N_A \rho \phi / M$: number of chains per volume) and $L = M \ell_0 / M_0$ a functional relation between the plateau modulus and the contour length density should hold

$$\frac{G_N^0 \ell_K^3}{k_B T} = F(vL\ell_K^2) \tag{51}$$

where the dimensional quantities G_N^0 and vL were made dimensionless by multiplying by ℓ_K and dividing by $k_B T$. If we further consider that $vL\ell_K^2$ is proportional to the polymer volume fraction ϕ and that in concentrated solutions the experimental result $G_N^0 \approx \phi^a$ with a $\cong 2$ holds, Eq. (51) assumes the form of a power law. If we use the relation of v, L, ℓ_K with molecular quantities, we obtain:

$$G_N^0 \approx k_B T C_\infty^{2a-3} \left(\frac{\rho\phi}{M_0}\right)^a \ell_0^{3a-3} \tag{52}$$

In the Doi-Edwards theory the plateau modulus and the tube diameter are related according to Eq. (40). Inserting Eq. (40) into (52) we finally obtain

$$d^2 \approx C_\infty^{4-2a} \left(\frac{\rho\phi}{M_0}\right)^{1-a} \ell_0^{5-3a} \tag{53}$$

This relation is the central result of the scaling model for the entanglement distance.

In the packing model [50, 62, 68] the entanglement distance is interpreted by the gradual build-up of geometrical restrictions due to the existence of other chains in the environment or, more precisely, the entanglement distance is determined by a volume which must contain a defined number of different chains. This approach is based on the observation that, for many polymer chains, the product of the density of the chain sections between entanglements is

$$n_e = \rho\phi N_A / M_e \tag{54}$$

(M_e: molecular mass of these entanglement strands) and the volume spanned by the entanglement distance is approximately a constant

$$\text{const} = n_e d^3 = \frac{\rho\phi N_A}{M_e} \left(C_\infty \frac{M_e}{M_0} \ell_0^2\right)^{3/2} \tag{55}$$

Neglecting the influence of chain ends, Eq. (55) yields for the entanglement distance

$$d^2 \approx \frac{1}{\left(\frac{\rho\phi}{M_0}\right)^2 C_\infty^2 \ell_0^4} \tag{56}$$

which is the special case of Eq. (53) for $a = 3$. We note that, as a consequence of the packing criterion, the entanglement distance increases with the rising tendency of a chain to coil; coiling reduces the presence of other chains in the volume spanned by a chain and leads to an increase in the entanglement volume.

A long-discussed alternative to the packing approach is the concept that a fixed number of binary contacts of different chains is required for an effective entanglement [63, 64]. This argument is again of a topological nature, based on the inability of chains to intersect each other. Let $\varphi_K = N_A \rho\phi/M_0 C_\infty$ be the density of Kuhn segments, then the number n_b of binary contacts/volume is given by

$$n_b \approx \varphi_K^2 \ell_K^3 \tag{57}$$

With the number of entanglement strands n_e/volume, the desired number of binary contacts per entanglement strand becomes

$$\frac{n_b}{n_e} = \text{const} \approx \left(\frac{\rho\phi}{m_0}\right) \ell_0^3 C_\infty N_e \tag{58}$$

As before, using Eq. (40) for the entanglement distance,

$$d^2 \approx \frac{1}{\left(\frac{\rho\phi}{m_0}\right) \ell_0} \tag{59}$$

is obtained.

The binary contact model is the special case of the scaling model [Eq. (53)] for $a = 2$.

The reason why simple scaling considerations do not lead to a unique result for the exponent 'a' is due to the fact that the entanglement problem as a geometrical phenomenon contains two independent lengths: the lateral distance between the chains $s = (L/V)^{-1/2}$ and the step width of the random walk ℓ_K. Further arguments similar to those above are therefore needed in order to define the exponent.

The Topological Approach. Like the scaling approach, the topological calculations are based on the assumption that the entanglement problem is purely geometrical, i.e. a mathematical problem, where dynamic effects do not play a role. The topological calculations go beyond scaling in that the topological invariants of the geometrical constraints are calculated rather than conjectured as in the scaling arguments. As entanglements are considered to be a purely geometrical effect, the entanglement distance can only be a function of the two length scales in the problem: the step length of the random walk of the chains ℓ_K and the distance between chain contours $(L/V)^{-1/2}$. Using the Gaussian topological invariants which count the windings swept out by one curve around another, Edwards [65] calculated the dependence of the tube diameter on the chain contour density for two limiting cases. For pure random flight polymers,

$$d \approx \frac{1}{\ell_K \left(\dfrac{L}{V}\right)} \approx \frac{1}{\left(\dfrac{\rho\phi\ell_0}{m_0}\right)\ell_K} \tag{60}$$

is obtained. This result agrees with a scaling argument given earlier by Doi [68]. He argued that for a chain undergoing Gaussian statistics, with the contour length density being the only physical quantity, the tube diameter should obey a scaling law

$$d \approx \left(\frac{L}{V}\right)^x \ell_L^{2x+1} \tag{61}$$

Since Gaussian statistics contain only L and ℓ_K in the combination $\langle R_e^2 \rangle = \ell_K L$, x has to be -1 which leads to Eq. (60). Incidentally, Eq. (60) corresponds to the Graessley-Edwards scaling formula for $a = 3$ and is also followed by the packing models.

The other limiting case concerns locally smooth chains. For the case of a Gaussian chain in a network of rods, Edwards found $d \approx (L/V)^{-1/2}$ [66] which agrees with the above discussed binary contact model. Finally, considering worm-like chain bridges the differences in the power laws between the two limiting cases and exponents between -1 and $-1/2$ may be obtained.

3.4.2 NSE Experiments on Entanglement Formation

An experimental test of the scaling model requires a selective variation of the two scaling variables of the model, i.e. the lateral chain distance and the chain stiffness. The Kuhn length ℓ_K depends on temperature via the characteristic ratio C_∞; the lateral chain distance s can be varied via the volume fraction ϕ.

On saturated PB (PEB-2; $M_W = 73\,200$ g/mol; $M_W/M_n = 1.02$), the dependence on contour length density was studied by diluting the melt with the short oligomer n-$C_{19}D_{40}$ [60]. Thus, a concentration range $0.25 \leqslant \phi \leqslant 1$ was covered. This system was mixed with a small fraction of protonated long chains so that it was again possible to study the dynamic structure factor of a single long chain. Figure 28 shows dynamic structure factors of such diluted melts with different paraffin content in the scaling plot of the Rouse model. In all cases, Q-dependent splitting of the spectra is observed which is characteristic for the appearance of the intermediary length scale or the tube diameter. Even in the case of the sample already containing 70% paraffin, splitting is observed after a long common course – Rouse motion takes place here. In total, polyethylene

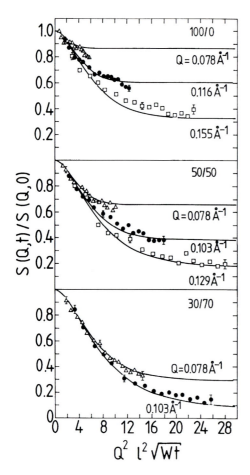

Fig. 28. NSE spectra in polyethylene melts at 509 K for three different polymer volume fractions in Rouse scaling. *Upper diagram* $\Phi = 1$, *central diagram* $\Phi = 0.5$, *lower diagram* $\Phi = 0.3$. The *solid lines* correspond to a fit with the Ronca model. (Reprinted with permission from [60]. Copyright 1993 American Chemical Society, Washington)

melts with seven different paraffin contents were examined. The data were analyzed with both the generalized Rouse model according to Ronca [50] and the local reptation model developed by de Gennes [53]. Figure 29 shows the resultant tube diameter as a function of the polymer volume fraction in a double logarithmic plot. First, it is to be noted that both models provide comparable results except for the lowest concentration. The tube diameter d follows a power law with $d \propto \phi^{-0.6}$. This behavior clearly differs from the prediction of the packing model Eq. (50) $d \propto \phi^{-1}$ and is close to the prediction of the binary contact model predicting $d \propto \phi^{-0.5}$. The value of the exponent a = 2.2 is also in good agreement with typical rheological results in concentrated polymer solutions.

Boothroyd et al. [74] recently determined the temperature dependence of the Kuhn length for polyethylene with the aid of small-angle neutron scattering. In the temperature range between 100 and 200 °C, $d \ln C_{\infty}/dT = -1.1 \times 10^{-3} \, K^{-1}$

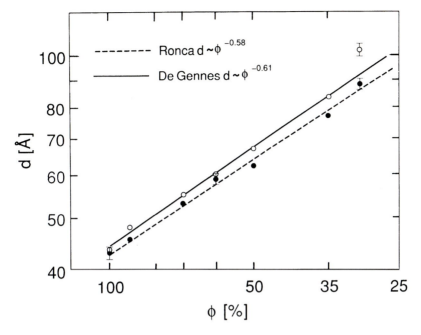

Fig. 29. Double logarithmic representation of the entanglement distance in polyethylene at 509 K as a function of the polymer volume fraction Φ. (Reprinted with permission from [60]. Copyright 1993 American Chemical Society, Washington)

is obtained for the temperature coefficient. Consequently, ℓ_K can be varied by about 20% over an experimentally accessible temperature range of 150 °C.

In the temperature range between 400 and 550 K, NSE spectra on the same undiluted polyethylene melt were recorded. These data were analyzed with respect to the entanglement distance. The result for the temperature-dependent entanglement distance d(T) is shown in Fig. 30. An increase in the tube diameter from about 38 to 44 Å with rising temperature is found.

Since both the temperature dependence of the characteristic ratio and that of the density are known, the prediction of the scaling model for the temperature dependence of the tube diameter can be calculated using Eq. (53); the exponent a = 2.2 is known from the measurement of the φ-dependence. The solid line in Fig. 30 represents this prediction. The predicted temperature coefficient $0.67 \pm 0.1 \times 10^{-3}\,K^{-1}$ differs from the measured value of $1.2 \pm 0.1 \times 10^{-3}\,K^{-1}$. The discrepancy between the two values appears to be beyond the error bounds. Apparently, the scaling model, which covers only geometrical relations, is not in a position to simultaneously describe the dependences of the entanglement distance on the volume fraction or the flexibility. This may suggest that collective dynamic processes could also be responsible for the formation of the localization tube in addition to the purely geometric interactions.

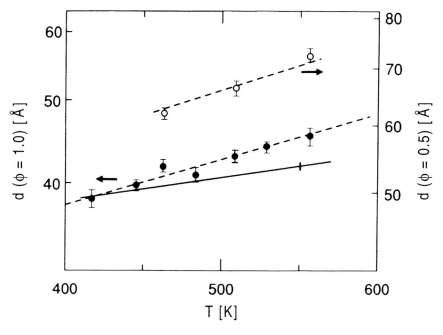

Fig. 30. Temperature dependence of the entanglement distance for polyethylene. ● Φ = 1; ○ Φ = 0.5. The *dotted lines* give a best fit for the data. The *solid line* represents the prediction by the scaling model of Graessley and Edwards (see text). (Reprinted with permission from [60]. Copyright 1993 American Chemical Society, Washington)

Finally, a further unsolved problem should be mentioned. If we compare the plateau moduli of different polymer melts and relate them to the Kuhn length and to the density, this relation can also be adequately described with the scaling model, if an exponent 'a' near 3 is chosen [73]. It is not known why this exponent is different if the contour length density is varied by dilution in concentrated solution or by selecting polymer chains of different volume.

In conclusion, scaling models link the formation of entanglements with geometric interactions between long polymer chains which cannot intersect one another. With the aid of a variation in temperature and concentration it was possible to vary the two parameters of the scaling model, the lateral chain distance s and the Kuhn length ℓ_K. The concentration-dependent measurements are in agreement with the binary contact model and contradict the packing model which, on the other hand, adequately describes the comparison of different polymer melts. The common dependence of the entanglement distance on the Kuhn length and the lateral chain distance leads to discrepancies from the scaling model.

4 Polymer Networks

4.1 Theoretical Outline – Dynamics of Junctions and Trapped Linear Chains

The macroscopic long-time behavior of dense polymer liquids exhibits drastic changes if permanent cross-links are introduced in the system [75–77]. Due to the presence of junctions the flow properties are suppressed and the viscoelastic liquid is transformed into a viscoelastic solid. This is contrary to the short-time behavior, which appears very similar in non-cross-linked and crosslinked polymer systems.

According to the importance of the cross-links, various models have been used to develop a microscopic theory of rubber elasticity [78–83]. These models mainly differ with respect to the space accessible for the junctions to fluctuate around their average positions. Maximum spatial freedom is warranted in the so-called phantom network model [78, 79, 83]. Here, freely intersecting chains and forces acting only on pairs of junctions are assumed. Under stress the average positions of the junctions are affinely deformed without changing the extent of the spatial fluctuations. The width of their Gaussian distribution is predicted to be

$$\langle d^2 \rangle = \frac{f-1}{f(f-2)} \langle R_e^2 \rangle \tag{62}$$

where f is the functionality of the network and $\langle R_e^2 \rangle$ the mean square end-to-end distance between topologically neighboring cross-links.

Assuming that the average positions of the junctions are uncorrelated and that Rouse dynamics prevail on short-time scales, the scattering function of the cross-links can be approximated by [84].

$$\frac{S(Q,t)}{S(Q,0)} = \exp\left\{-\frac{Q^2\langle d^2\rangle}{4}\right\} + \left[1 - \exp\left\{-\frac{Q^2\langle d^2\rangle}{4}\right\}\right]$$

$$\times \exp\left\{-\frac{12}{\sqrt{\pi}}(\Omega(Q)t)^{1/2}\right\} \tag{63}$$

According to Warner [85], $\Omega(Q)$ is related to the characteristic frequency $\Omega_R(Q)$ of linear chains [see Eq. (22)] by

$$\Omega(Q) = \frac{2}{f}\Omega_R(Q) \tag{64}$$

The term

$$F(Q) = \exp - \{Q^2\langle d^2\rangle/4\} \tag{65}$$

which is the Fourier transform of $\exp\{-r^2/\langle d^2\rangle\}$, results from the fact that there is a finite probability to find the junction at a position r close to its initial position, even at long times. In the language of quasi-elastic neutron scattering, this contribution is called elastic incoherent structure factor (EISF) [29]. It should be mentioned that the scattering of the cross-links, although being coherent, can be described by the incoherent dynamic structure factor, as long as their positions and dynamics are uncorrelated.

If long linear polymer chains are trapped in a network of strands, which are chemically identical to the unattached chains, the question arises whether the junctions as permanent entanglement induce a similar relaxation as the non-permanent entanglements in melts of corresponding long linear homopolymer chains. In principle with these systems it is easy to vary the distance between topologically neighbored entanglements to a larger extent and to analyze the chain dynamics as a function of this parameter.

4.2 NSE Results from Polymer Networks

Dynamics of Junctions

NSE measurements on the junction dynamics were performed on four-functional PDMS model networks at $T = 373$ K [84]. They were prepared by end-linking deuterated bifunctional precursor PDMS chains ($M_W^D = 5500$ g/mol, $M_W^D/M_n^D = 1.8$) with four-functional protonated cross-links, each of them containing 22 protons. For comparison, a melt of the corresponding end-labelled precursor polymers was also investigated. In order to look for correlations between the cross-links, SANS experiments were carried out, too. In the Q-range $0.001 < Q < 0.3\,\text{Å}^{-1}$, the intensity profile exhibits nearly no Q-dependence. A slight decay towards larger Q is probably due to the form factor of the individual cross-links which have a radius of gyration of about $10\,\text{Å}$. Since correlations between different cross-links were not observed, their scattering can be treated as originating from a single junction, and it reveals the self-correlation function of the cross-links.

The mesh size was determined from an equivalent network, prepared from a mixture of protonated and deuterated precursors. The radius of gyration of the mesh strands was found to be 25 Å, which is identical to the radius of gyration of the uncross-linked precursors.

Figure 31 compares the dynamic structure factors obtained from the cross-links and the chain ends for two different Q-values. Without any analysis a strong reduction of the cross-link mobility compared to that of the chain end is obvious. A closer inspection also shows that the line-shape of both curves differs. While $S(Q, t)/S(Q, 0)$ from the chain end decays continuously, $S(Q, t)$ from the cross-links appears to decay faster at shorter than at longer times. This difference in line shape is quantified via the line shape parameter β. For the end-labelled chains, β is in close agreement with the $\beta = 1/2$ prediction of the

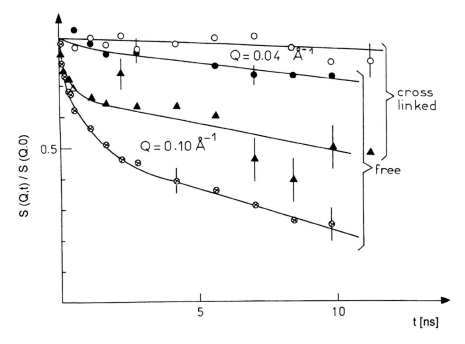

Fig. 31. NSE spectra of linear end-labelled PDMS chains (M_w = 5500 g/mol) and of four-functional cross-links at T = 373 K. The *solid lines* are visual aids. (Reprinted with permission from [84]. Copyright 1988 The American Physical Society, Maryland)

self-correlation function of a Rouse chain [see Eq. (21)]. The experimentally determined $W\ell^4 = 3.2\ 10^{13}\ \text{Å}^4\,\text{s}^{-1}$ exceeds the corresponding value of linear PDMS chains, for which $W\ell^4 = 1.7\ 10^{13}\ \text{Å}^4\,\text{s}^{-1}$ is measured, by a factor of nearly two, as expected from Eq. (64). The observation of an initially faster decaying relaxation curve for the cross-links translates into a smaller exponent β with a mean value of 0.40 ± 0.04.

Figure 32 presents the fit of Eq. (63) to the cross-link data. The shaded area displays the constant EISF contribution to S(Q, t). A combined fit to all spectra yields $W\ell^4 = 0.84\ 10^{13}\ \text{Å}^4\,\text{s}^{-1}$ and $\langle d^2 \rangle^{1/2} = 24.5 ± 1.5\ \text{Å}$. Displayed are fits to single spectra with the rate fixed to the value obtained from the joint fit. The obtained $\langle d^2 \rangle^{1/2}$-values vary between 22.6 and 28.1 Å and demonstrate the consistency of the description. The obtained Rouse parameter amounts to nearly exactly half the value of the high molecular mass melt of linear PDMS chains in very good agreement with the prediction of a 2/f reduction [see Eq. (64)].

These measurements for the first time allowed experimental access to the microscopic extent of cross-link fluctuations. The observed range of fluctuation is smaller than predicted by the phantom network model, for which

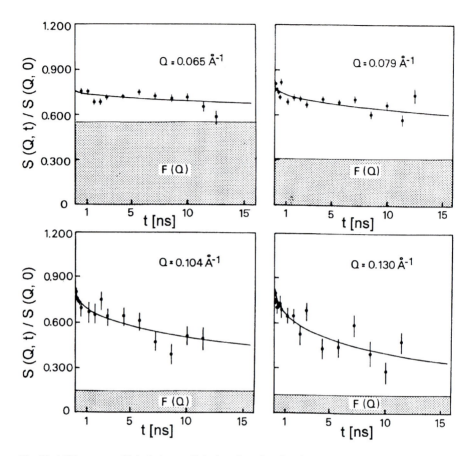

Fig. 32. NSE spectra of labelled cross-links in a four-functional PDMS network at four different Q-values. Included is a common fit with Eq. (63). The *shaded area* displays the time independent EISF part in the spectra. Note that the spectra do not approach 1 for $t \to 0$. This is related to a fast relaxation process of the deuterated network strands which has not been substracted. (Reprinted with permission from [84]. Copyright 1988 The American Physical Society, Maryland)

$\langle d^2 \rangle^{1/2} = 37.5\,\text{Å}$ is expected on the basis of Eq. (62), when $\langle R_e^2 \rangle$ is replaced by $6\langle R_g^2 \rangle$ and $\langle R_g \rangle^{1/2} = 25\,\text{Å}$ is used. These deviations suggest that in a real network additional constraints are active, which reduce the range of fluctuations below the phantom network prediction.

A model, which accounts for this effect, is the junction constraint model of Flory [86]. Starting from the phantom network an additional parameter κ

$$\kappa = d^2_{phantom}/d^2_{obs} - 1 \tag{66}$$

is introduced to represent the degree of constraints operating on the junctions [87]. κ is assumed to be proportional to the number of junctions in the volume pervaded by a network chain and can be expressed in terms of molecular parameters as [88].

$$\kappa = \frac{1}{f} N_A \rho (C_\infty \ell^2 / M_0)^{3/2} M_S^{1/2} \tag{67}$$

(M_S, M_0 molecular mass per strand and monomer, respectively).

Whereas $\kappa = 1.3$ is derived from the above-presented NSE data, $\kappa = 2.75$ is expected for a four-functional PDMS network of $M_S = 5500$ g/mol on the basis of Eq. (67). Similar discrepancies were observed for a PDMS network under uniaxial deformation [88]. However, in reality this discrepancy may be smaller, since Eq. (67) provides the upper limit for κ, calculated under the assumption that the network is not swollen during the cross-linking process due to unreacted, extractable material. Regardless of this uncertainty, the NSE data indicate that the experimentally observed fluctuation range of the cross-links is underestimated by the junction constraint and overestimated by the phantom network model [89].

Linear Chains Trapped in Networks

The influence of the network junctions on the dynamics of high molecular mass linear PDMS chains ($M_W = 10\,000$ g/mol) was studied on two different PDMS model networks with mesh sizes corresponding to $M_W = 5500$ and 2700 g/mol, respectively. For comparison, a melt and a concentrated solution (c = 0.6) of corresponding linear PDMS chains were also investigated [84]. All measurements were performed at T = 313 K, using the NSE spectrometer IN11 with a time window limited to 12 ns. The line-shape parameters β, as derived from the different samples, have a value of 0.5 within experimental error, indicating that Rouse dynamics prevail in the time regime experimentally probed. In Fig. 33 the corresponding characteristic frequencies $\Omega_R(Q)$ are plotted vs. Q on a double logarithmic scale. They were determined by fitting the coherent scattering function of the Rouse model (see Table 1) to the respective spectra. The solid lines display the $\Omega_R(Q) \sim Q^4$ scaling law of the Rouse model. In all cases this relationship is followed very closely by the data. One realizes further that with increasing constraints the curves are shifted towards lower rates, indicating a reduction of the Rouse parameter with increasing hindrance. Going from the concentrated solution to the trapped chain in a network with only about 40 monomers between cross-links, the entanglement distance is changed by a factor 2.5 and brought into the Q-range of the experimental setup. Despite this important change in the constraints acting on a chain, the experimental data do not change their characteristics either with respect to the line shape or with respect to the Q-scaling of the characteristic frequency on this limited time scale. While the difference between the melt and the concentrated solution is easily attributed to a changing segmental friction coefficient, reduced to the addition of solvent, the difference between both network systems is quite unexpected, since

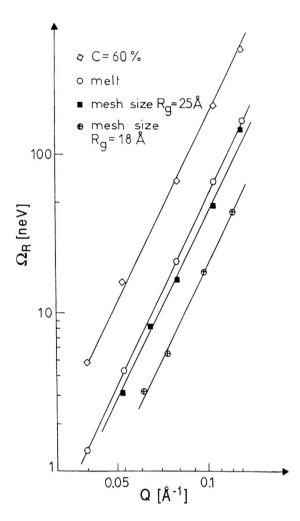

Fig. 33. Characteristic frequencies for the concentrated solution, the PDMS melt and trapped chains. The *solid lines* display the $\Omega_R \sim Q^4$ scaling law. (Reprinted with permission from [13]. Copyright 1988 Mansgaard International Publishers Ltd., Copenhagen)

the friction coefficient as a local property should not change on corresponding melt and network systems.

5 Polymer Dynamics in the Presence of a Solvent

The dynamics of polymer solution is governed by the hydrodynamic interaction between moving segments mediated by the solvent [90, 91]. This interaction is long range and couples the motions of the different segments strongly. In this section we discuss first the impact of this hydrodynamic interaction in terms of the so-called Zimm model [7] which generalized the Rouse model for the presence of hydrodynamic coupling and then calculate the dynamic structure factor for this model. Thereafter experimental results on dilute polymer solutions are presented, addressing the case of good and Θ solvents as well as the crossover between both conditions.

Next, we consider dilute solutions of non-linear polymers and report on the dynamics of ring- and star-shaped macromolecules. Then we move to semi-dilute solutions, where interaction effects between chains become important, giving rise to collective dynamics of the many chain systems. For linear homopolymers the crossover properties from the single to collective are discussed in particular. Considering the dynamics of single chains within a semi-dilute solution, with rising concentration the hydrodynamic interactions become increasingly screened, driving a transition from Zimm dynamics at smaller scales to Rouse dynamics at larger scales. Experimental results indicate a further so-called intermediate Zimm regime. Finally, we turn again from linear homopolymers to polymers with different architectures and review the results on semi-dilute solution of star-shaped and linear block copolymers which reveal a breathing mode of the two blocks moving with respect to one another.

5.1 Dilute Solution of Linear Polymers

5.1.1 Theoretical Outline – The Zimm Model

In addition to the influence of entropic forces treated in the Rouse model (Sect. 3.1), we now consider hydrodynamic interactions between different segments which are immersed in a solvent. A segment moving relative to the surrounding solvent generates a flow field which influences the movement of all other segments. With the aid of the Oseen tensor $\underline{\underline{T}}$ it is possible to specify which velocity is generated in a liquid at location $\underline{r}$ if the force $\underline{F}$ acts on a volume element at the origin

$$v(\underline{r}) = \underline{\underline{T}}\,\underline{F} \tag{68}$$

with

$$\underline{\underline{T}}_{nm} = \frac{1}{8\pi\eta_s|r_{nm}|}\left(\underline{\underline{E}} - \frac{\underline{r}_{nm} \oplus \underline{r}_{nm}}{r_{nm}^2}\right) \quad n \neq m \tag{69}$$

where, $\underline{r}_{nm} = \underline{r}_n - \underline{r}_m$, $\underline{\underline{E}}$ is the identity matrix, $\oplus$ the outer product and η_s the viscosity of the solvent. For illustration, Fig. 34 presents schematically the flow field resulting from a force in the y-direction.

In the equation of motion (13), the friction force with the solvent

$$\underline{F}_n = \zeta\underline{v}(r_n) \tag{70}$$

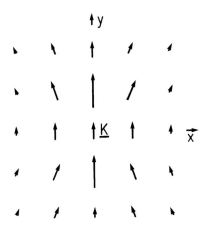

Fig. 34. Schematic two-dimensional representation of the flow field which is generated by a force at the origin in the y-direction. The *arrow length* and *directions* symbolize the velocities of the solvent. (Reprinted with permission from [14]. Copyright 1992 Kluwer Academic Publishers, Dordrecht)

is now added to the entropic force. The flow field $v(\underline{r}_n)$ is the result of all forces exercised on the solvent by all other segments

$$\underline{v}(\underline{r}_n) = \sum_{n \neq m} \underline{\underline{T}}_{nm}\left(k_e \frac{\partial^2 \underline{r}_m}{\partial m^2} + \underline{f}_m(t)\right) \tag{71}$$

We can take the Rouse term $1/\zeta k_e \, \partial^2 r_m/\partial m^2$ ($k_e = 3k_B T/\ell^2$): entropic spring constant) into consideration formally, if we define the element $\underline{\underline{T}}_{nm}$ of the Oseen tensor as $\underline{\underline{T}}_{nm} = \underline{\underline{E}}/\zeta$. The equation of motion (13) thus becomes

$$\frac{\partial \underline{r}_n}{\partial t} = \sum_m \underline{\underline{T}}_{nm}\left(k_e \frac{\partial^2 \underline{r}_m}{\partial m^2} + \underline{f}_m(t)\right) \tag{72}$$

Equation (72) is a coupled non-linear differential equation which is difficult to handle. We therefore consider a number of approximations.
1. We average Eq. (69) over all possible orientations

$$\langle \underline{\underline{T}}_{nm}\rangle_\Omega = \underline{\underline{E}} \frac{1}{6\pi\eta_s |\underline{r}_n - \underline{r}_m|} \tag{73}$$

2. For small deviations from equilibrium, we average $|\underline{r} - \underline{r}_m|^{-1}$ by the statistical segment distribution. For a Gaussian chain which is realized under Θ-conditions we obtain

$$\langle \underline{\underline{T}}_{nm}\rangle_{eq} = \underline{\underline{E}} \int_0^\infty dr \, 4\pi r^2 \left(\frac{3}{2\pi|n-m|\ell^2}\right)^{3/2} \exp\left(-\frac{3r^2}{2|n-m|\ell^2}\right)\frac{1}{6\pi\eta_s r}$$

$$= \frac{\underline{\underline{E}}}{6\pi^3 |n-m|^{1/2}\eta_s \ell} \tag{74}$$

With the aid of these approximations, (72) is linearized

$$\frac{\partial r_n(t)}{\partial t} = W \frac{\partial r_n(t)}{\partial n^2} + f_n(t) + W \left(B \sum_{n \neq m} \frac{1}{|n-m|^{1/2}} + f_m(t) \right) \tag{75}$$

where W is the elementary relaxation rate of the Rouse process [see Eq. (16)] and

$$B = \frac{1}{6^{1/2} \pi^{3/2}} \frac{\zeta}{\eta_s \ell} \tag{76}$$

the draining parameter, by which the strength of hydrodynamic interaction is taken into account. B = 0 is the Rouse limit, where no hydrodynamic interaction is present; whereas B = 0.4 is the Zimm limit, where hydrodynamic interaction is dominant.

The long-range coupling via the flow field which only decreases with $1/r$ leads to a qualitatively different behavior from that of the Rouse model. Equation (75) is approximately solved by transformation to Rouse normal coordinates. Its solution [6, 91] leads to the spectrum of relaxation rates

$$\frac{1}{\tau_p} = W \left(1 - \cos \frac{\pi p}{N} \right) \left(2 + 4B \sum_{s=1}^{\infty} s^{-1/2} \cos \frac{\pi p s}{N} \right) \quad p = 1, 2, \ldots, N \tag{77}$$

with $s = |n-m|$.

Restricting the low frequency part of the spectrum, which occurs for $p \ll N$, this result simplifies to

$$\frac{1}{\tau_p} = W \left(\frac{\pi}{N} p \right)^2 \left(1 + (2\pi)^{1/2} B \left(\frac{p\pi}{N} \right)^{-1/2} \right) \quad p = 1, 2, \ldots, N \tag{78}$$

Although at first sight the approximations used to derive (72) seem to be very rough, detailed calculations nevertheless show that the results deviate only very slightly from the exact solutions [93, 94].

In contrast to the Rouse modes, which have the dispersion $\tau_p^{-1} \sim p^2$, the pure Zimm modes (B ≠ 0 and $B\sqrt{2N/p} \gg 1$) lead to

$$\frac{1}{\tau_p} = \frac{1}{\tau_z} p^{3/2} \tag{79}$$

with

$$\tau_z = \frac{\eta_s N^{3/2} \ell^3}{\sqrt{3\pi k_B T}} = 0.325 \frac{\eta_s R_e^3}{k_B T} \tag{80}$$

In addition, analogously to (20), for $t < \tau_z$ the mean square displacement of a diffusing polymer segment becomes

$$\Delta r_{nn}^2 = \frac{16}{9} \frac{1}{\pi^{3/2}} \left(\frac{\sqrt{3\pi k_B T}}{\eta_s} t \right)^{2/3} \tag{81}$$

Table 4. Assertions of the Zimm and Rouse model on molecular dynamics

Property	Zimm	Rouse
Center of mass diffusion coefficient	$0.196 \dfrac{k_B T}{\eta_s R_e}$	$\dfrac{k_B T}{N\zeta}$
Longest relaxation time	$\tau_z = 0.325 \dfrac{\eta_s R_e^3}{k_B T}$	$\tau_R = \dfrac{\zeta R_e^4}{3\pi^2 k_B T \ell^2}$
Eigenvalue spectra	$\approx \dfrac{1}{p^{3/2}}$	$\approx \dfrac{1}{p^2}$
Segmental mean square displacement	$\dfrac{16}{9\pi^{3/2}}\left(\dfrac{\sqrt{3\pi}k_B T}{\eta_s}t\right)^{2/3}$	$2\ell^2\left(\dfrac{3k_B T}{\pi\zeta\ell^2}t\right)^{1/2}$
Parameter	η_s	$\dfrac{\zeta}{\ell^2}$

Moreover, the center of mass diffusion is given by

$$D_z = \frac{8k_B T}{3(6\pi^3)^{1/2}\,\eta_s\ell\sqrt{N}} = 0.196\,\frac{k_B T}{\eta_s R_e} \tag{82}$$

In fact, the diffusion constant in solutions has the form of an Einstein diffusion of hard spheres with radius $\approx R_e$. For a diffusing chain the solvent within the coil is apparently also set in motion and does not contribute to the friction. Thus, the long-range hydrodynamic interactions lead, in comparison to the Rouse model, to qualitatively different results for both the center-of-mass diffusion—which is not proportional to the number of monomers exerting friction – as well as for the segment diffusion – which is considerably accelerated and follows a modified time law $\langle\Delta r^2\rangle \approx t^{2/3}$ instead of $t^{1/2}$.

The equations of motion (75) can also be solved for polymers in good solvents. Averaging the Oseen tensor over the equilibrium segment distribution then gives $\langle 1/\underline{r}_n - \underline{r}_m\rangle = 1/|n-m|^\nu\ell$; $\tau_p^{-1} = p^{3\nu}/\tau_z$ and $D_z \approx k_B T/\eta_s N^\nu\ell$ are obtained for the relaxation times and the diffusion constant. The same relations as (80) and (82) follow as a function of the end-to-end distance with slightly altered numerical factors. In the same way, a solution of equations of motion (75), without any orientational averaging of the hydrodynamic field, merely leads to slightly modified numerical factors [35]. In conclusion, Table 4 summarizes the essential assertions for the Zimm and Rouse model and compares them.

Dynamic Structure Factors for the Zimm Model

As in the case of the Rouse dynamics (see Sect. 3.1.1), the intermediate incoherent scattering law for dominant hydrodynamic interaction (Zimm model) can be obtained directly from Eqs. (4b) and (81)

$$\frac{S_{inc}(Q,t)}{S_{inc}(Q,0)} = \exp - \left\{ \frac{2^{11/2}}{9\sqrt{\pi}} (\Omega_Z(Q)t)^{2/3} \right\} \tag{83}$$

with

$$\Omega_Z(Q) = \frac{\sqrt{6\pi}}{18} W_R B\ell^3 Q^3 = \frac{1}{6\pi} \frac{k_B T}{\eta_s} Q^3 \tag{84}$$

the related characteristic frequency.

Comparing Eqs. (83), (84) and Eqs. (21), (22) it follows immediately that Rouse and Zimm relaxation result in completely different incoherent quasi-elastic scattering. These differences are revealed in the line shape of the dynamic structure factor or in the β-parameter if Eq. (23) is applied, as well as in the structure and Q-dependence of the characteristic frequency. In the case of dominant hydrodynamic interaction, $\Omega(Q)$ depends on the viscosity of the pure solvent, but on no molecular parameters and varies with the *third* power of Q, whereas with failing hydrodynamic interaction it is determined by the inverse of the friction per mean square segment length and varies with the *fourth* power of Q.

The coherent structure factor of the Zimm model can be calculated [34] following the lines outlined in detail in connection with the Rouse model. As the incoherent structure factor (83), it is also a function of the scaling variable $(\Omega_Z(Q)t)^{2/3}$ and has the form

$$S(Q,t) = \frac{12}{Q^2\ell^2} \int_0^\infty du \exp\{ -u - (\Omega_Z t)^{2/3} g(u(\Omega_Z t)^{2/3}) \} \tag{85}$$

with

$$g(y) = \frac{2}{\pi} \int_0^\infty dx \frac{\cos xy}{x^2} (1 - \exp(-2^{-1/2}x^{2/3})) \tag{86}$$

From Fig. 35, where the normalized coherent scattering laws $S(Q,t)/S(Q,0)$ are plotted as a function of $\Omega(Q)t$ for Zimm as well as for Rouse relaxation, one sees that hydrodynamic interaction results in a much faster decay of the dynamic structure factor.

The long-time behavior $(\Omega(Q)t) > 1$ of the coherent dynamic structure factors for both relaxations shows the same time dependence as the corresponding incoherent ones

$$\frac{S(Q,t)}{S(Q,0)} = \exp\left\{ -\frac{2}{\sqrt{2}} (\Omega_R t)^{1/2} \right\} \tag{87}$$

for the Rouse relaxation and

$$\frac{S(Q,t)}{S(Q,0)} = \exp\left\{ -\frac{2^{2/3}}{\pi} \Gamma(1/3) (\Omega_Z t)^{2/3} \right\} \tag{88}$$

for the Zimm relaxation, where $\Gamma(x)$ is the Γ-function.

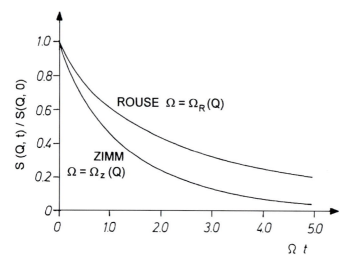

Fig. 35. Dynamic structure factors $S(Q, t)/S(Q, 0)$ as dependent on Ωt for the Rouse and the Zimm model

For the case that both Rouse *and* Zimm relaxation are present, the coherent dynamic structure factor is also available [40]. For completeness, its analytical representation is included in Table 1.

Initial Slope or First Cumulant

More general access to the coherent dynamic structure factor was provided by Akcasu and coworkers [93–95], starting from the assumption that the temporal evolution of the densities in Fourier space $\rho(Q, t)$

$$\rho(\underline{Q}, t) = \sum_j \exp\{i\underline{Q}\,\underline{r}_j(t)\}, \tag{89}$$

being related to the coherent dynamic structure factor by

$$S(Q, t) = \langle \rho(Q, t)\rho^*(Q, 0)\rangle, \tag{90}$$

is governed by an equation of the form

$$\frac{\partial \rho(Q, t)}{\partial t} = -\hat{L}\rho(\underline{Q}, t) \tag{91}$$

where $\hat{L}$ is a real, linear and time-independent operator acting on the position vectors $r_j(t)$ of all monomers. (By * the complex conjugated and by $\langle \cdots \rangle$ the thermal average are indicated.) Making use of the Zwanzig-Mori projection operator formalism [96, 97], the following exact equation for $\rho(Q, t)$, often referred to as the "generalized Langevin equation" is derived from (91)

$$\frac{\partial \rho(Q, t)}{\partial t} = -\Omega(Q)\rho(q, t) + \int_0^t du\, K(Q, t - u)\, \rho(Q, t) + f(Q, t) \tag{92}$$

where $f(Q, t)$ is the stochastic force,

$$\Omega(Q) = \frac{\langle \rho(Q, 0)\hat{L}\rho^*(Q, 0)\rangle}{\langle \rho(Q, 0)\rho^*(Q, 0)\rangle} \tag{93}$$

the characteristic frequency in Fourier space and $K(Q, t)$ a memory kernel accounting for the non-Markovian nature of the time dependence of $S(Q, t)$. It should be pointed out that only static properties enter the calculation of $\Omega(Q)$, which can also be written in terms of a generalized mobility $\mu(Q)$ as [98]

$$\Omega(Q) = \frac{k_B T Q^2 \mu(Q)}{S(Q, 0)} \tag{94}$$

In the case of dilute polymer solutions, $\hat{L}$ is conventionally taken to be the Kirkwood-Riseman diffusion operator [99]. As a consequence, the generalized mobility $\mu(Q)$

$$\mu(Q) = \frac{1}{k_B T Q^2} \langle \rho(Q, 0)\hat{L}\rho^*(Q, 0)\rangle \tag{95}$$

can be expressed in terms of the Oseen diffusion tensor

$$\mu(Q) = \frac{1}{k_B T Q^2} \sum_{j, K} \langle Q \underline{\underline{D}}_{jK} Q \exp\{-iQr_{jK}\}\rangle \tag{96}$$

with

$$\underline{\underline{D}}_{jK} = \frac{k_B T}{\zeta} \delta_{jK} + (1 - \delta_{jk})\underline{\underline{T}}_{jK} \tag{97}$$

where $\underline{\underline{T}}_{jK}$ is given by (69).

Due to the tensor character of $\underline{\underline{D}}_{jK}$ the thermal average is Eq. (96) leads to bulky expressions. However, it was shown [100] that (96) can be approximated at high accuracy by double sums of pre-averaged terms

$$\mu_{jK}^{pre}(Q) = \frac{1}{Q^2 k_B T} \langle Q \underline{\underline{D}}_{jK} Q\rangle \langle \exp\{iQr_{jK}\}\rangle \tag{98}$$

and correction terms $\Delta\mu_{jK}(Q)$.

Thus, finally

$$\mu^{pre}(Q) = \frac{1}{3\pi^2 \eta_s} \int_0^\infty S(Q^2 + u^2)^{1/2}\, du \tag{99}$$

and

$$\Delta\mu(Q) = \frac{Q^2}{30\pi^2 \eta_s} \int_0^\infty \frac{1}{u^2} [S(0.72Q^2)^{1/2} - S(0.72Q^2 + u^2)^{1/2}] \, du \qquad (100)$$

are obtained.

Since the random forces $f(Q, t)$ at time t are not correlated with the densities $\rho(Q, t')$ at $t \neq t'$, the time evolution of the dynamic structure factor (90) follows immediately from (92) as

$$\frac{\partial S(Q, t)}{\partial t} = -\Omega(Q)S(Q, t) + \int_0^t du \, K(Q, t - u) S(Q, u) \qquad (101)$$

If all memory effects are neglected, the solution of (101) is

$$S(Q, t) = S(Q, 0) \exp - \{\Omega(Q)t\} \qquad (102)$$

and $\Omega(Q)$ turns out to be the initial slope or first cumulant of $S(Q, t)$

$$\Omega(Q) = \lim_{t \to 0} \frac{\partial}{\partial t} \ln S(Q, t) \qquad (103)$$

Besides the fact that the prediction of $\Omega(Q)$ does not require detailed knowledge of the coherent dynamic structure factor $S(Q, t)/S(Q, 0)$, its calculation is neither restricted to the intermediate Q-range ($R_g Q \gg 1$; $\ell Q \ll 1$) (see Fig. 36) nor to

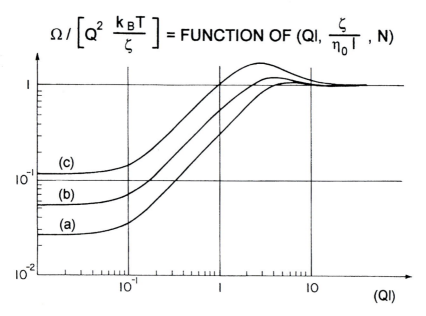

Fig. 36a–c. Plot of $\Omega/(Q^2(k_BT/\xi))$ vs. $Q\ell$ for $N = 10^3$. **a** $\xi/\eta_s\ell = 3\pi/2$; **b** $\xi/\eta_s\ell = 3\pi$; **c** $\xi/\eta_s\ell = 6\pi$. (Reprinted with permission from [93]. Copyright 1976 J Wiley and Sons, New York)

Θ-conditions. ($\langle R_g^2 \rangle$ is the mean square radius of gyration and R_g stands for $\langle R_g^2 \rangle^{1/2}$). In addition, it also works with a non-preaveraged Oseen tensor and is applicable to systems with finite chain length and chain-length distributions [100].

The application of Eq. (94) is very simple for Rouse relaxation. According to (96) and (97), $\mu(Q) = N/\zeta$ results. Due to the local character of friction, the mobility does not depend on Q. For $QR_g \ll 1$, where $S(Q,0) \simeq N^2$ is valid, $\Omega(Q) = k_B T Q^2/(N\zeta)$ the well-known center of mass diffusion behavior is obtained. In contrast, for $QR_g \gg 1$ and $Q\ell \ll 1$, where the internal modes are probed, $S(Q,0)$ is approximated by $2N/(Q\ell)^2$, and $\Omega(Q) \equiv \Omega_R = k_B T \ell^2 Q^4/(12\zeta)$ [see Eq. (22)] is derived.

In the case of Zimm dynamics the mobility is non-local due to the long-range hydrodynamic interactions. Being interested in the internal modes of Θ-systems, $\mu^{pre}(Q) = k_B T/(3\pi\eta_s Q R_g^2)$ results from Eq. (99) if $S(Q,0) = 12N/(Q\ell)^2$ is used. Thus, $\Omega(Q) \equiv \Omega_Z(Q) = k_B T Q^3/(6\pi\eta_s)$ [see Eq. (84)] is confirmed. If the Oseen tensor is not pre-averaged, the prefactor $1/(6\pi) = 0.053$ has to be replaced by 0.063 [93, 94].

Under good solvent conditions, where the asymptotic behavior of $S(Q,0) \sim Q^{-1/\nu}$ with $\nu = 2/3$ is valid in the regime $QR_g \gg 1$, and $Q\ell \ll 1$, Eq. (94) leads to $\Omega(Q) \sim (k_B T/\eta) Q^3$, too. However, the prefactors are larger than for Θ-conditions and have values of 0.071 and 0.079 for preaveraged and non-preaveraged Oseen tensors, respectively [100].

Scaling Analysis

Special theoretical insight into the internal relaxation behavior of polymers can also be provided on the basis of dynamic scaling laws [4, 5]. The predictions are, however, limited since only general functional relations without the corresponding numerical prefactors are obtained.

Within the scaling concept the answer to the question, how physical properties alter with changes in scale, is of fundamental importance. For example, Fig. 37 shows two polymer chains which result from another altering of the scale (here, number of segments N) by a factor $\lambda = 2$. Then the following transformation rules are valid for the monomer concentration c and the segment length ℓ [5]

$$N \rightarrow N/\lambda$$

$$c \rightarrow c/\lambda$$

$$\ell \rightarrow \ell \lambda^\nu \tag{104}$$

The final rule follows from the transformational invariance of the mean squared end-to-end distance $\langle R_e^2 \rangle = \ell^2 N^{2\nu}$ (ν Flory exponent). As a hypothesis it is

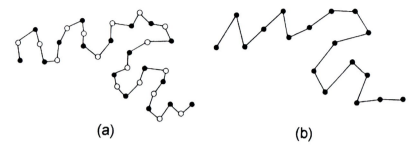

Fig. 37a, b. Two chains which result from each other by changing the segmental length scale by a factor $\lambda = 2$. **a** initial chain; **b** chain with altered scale.

assumed that each further physical quantity A, whether static or dynamic, is transformed by

$$A(\lambda x) = \lambda^P A(x) \tag{105}$$

Transformational invariant quantities lead to $p \equiv 0$.

On the basis of scaling arguments, general functional dependencies can also be derived. For example, dimensional analysis shows that the center of mass diffusion coefficient D_G for Zimm relaxation has the form

$$D_G = \frac{kT}{\eta_s \ell} f(N) \tag{106}$$

Since D_G has to be invariant under scaling transformations

$$\frac{k_B T}{\eta_s \ell} f(N) = \frac{k_B T}{\eta_s \ell \, \lambda^v} f(N/\lambda) \tag{107}$$

is true for arbitrary λ only for

$$f(N) \sim N^{-v} \tag{108}$$

This leads to $D_G \sim k_B T/(\eta_s R_e)$ (see Table 4).

To guarantee the transformational invariance of $D_G = k_B T/(N\zeta)$ (see Table 4) in the case of Rouse relaxation, the replacement of N by N/λ requires the simultaneous replacement of the segmental friction coefficient ζ by $\zeta \lambda$ which is natural, since friction is proportional to the number of segments involved.

Within the framework of the Rouse model the characteristic frequency for the center of mass diffusion follows the equation

$$\Omega(Q) = \frac{kT}{N\zeta} Q^2 \tag{109}$$

In order to include the internal diffusion, one has to start from

$$\Omega_R(Q) \simeq \frac{k_B T}{N\zeta} Q^2 f(QR_e) \simeq \frac{k_B T}{N\zeta} Q^2 (Q\ell N^{1/2})^x \tag{110}$$

where $f(Q, R_g)$ is the scaling function which usually is a power of the scaling variable. The exponent x follows from the requirement that for $QR_g \ll 1$, Eq. (110) has to be identical to Eq. (109) and that for $QR_g \gg 1$, Eq. (110) has to be independent of the chain length or the number of segments N. This leads to $x = 0$ in the case of center of mass diffusion and $x = 2$ and $\Omega_R(Q) \simeq (k_B T\ell^2/\zeta)Q^4$ [see Eq. (22)] in case of segmental diffusion.

For the Zimm model, the approach and argumentation are quite similar:

$$\Omega_Z(Q) \simeq \frac{k_B T Q^2}{\eta_s \ell N^{\nu_D}}(QR_e)^x \simeq \frac{k_B T Q^2}{\eta_s \ell N^{\nu_D}}(Ql N^\nu)^x \tag{111}$$

with $\nu = \nu_D = 1/2$ for Θ- and $\nu = \nu_D = 3/5$ for good solvent conditions. Thus, x has been set to 0 and 1 in the regimes $QR_g \ll 1$ and $QR_g \gg 1$, respectively. With respect to the segmental diffusion this leads to $\Omega_Z \simeq (k_B T/\eta_s)Q^3$ in the case of Θ- [see Eq. (84)] as well as of good solvent conditions.

Crossover from Θ- to Good Solvent Conditions

Based on the analogy between polymer solutions and magnetic systems [4, 101], static scaling considerations were also applied to develop a phase diagram, where the reduced temperature $\tau = (T - \Theta)/\Theta$ (Θ: Θ-temperature) and the monomer concentration c enter as variables [102, 103]. This phase diagram covers Θ- and good solvent conditions for dilute and semi-dilute solutions. The latter will be treated in detail below.

The relevant part of the phase diagram ($\tau > 0$) is shown in Fig. 38. The c-τ-plane is divided into four areas. The dilute regime I' and I are separated from the semi-dilute regimes III and II, where the different polymer coils interpenetrate each other, by the so-called overlap concentration

$$c^* = \frac{M}{N_A \langle R_g^2 \rangle^{3/2}} \tag{112}$$

(M: molar mass), where I' and III are the tricritical or Θ-regions. Here, the chain molecules exhibit an unperturbed random coil confirmation. In contrast, I and II are the critical or good solvent regimes, which are characterized by structural fluctuations in direction of an expanded coil conformation. According to the underlying concept of critical phenomena, the phase boundaries have to be considered as a continuous crossover and not as discontinuous transitions.

The above-mentioned conformational fluctuations can be easily understood on the basis of the "blob" model [104–109]. For example, this is illustrated for the transition from I to I'. Within the chain swollen by excluded-volume

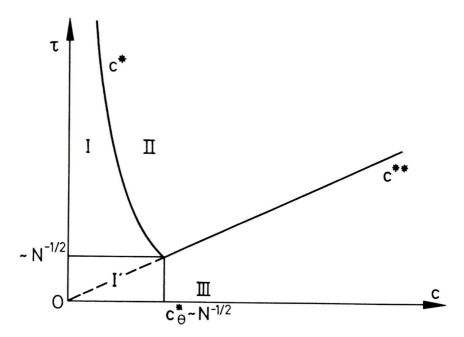

Fig. 38. Temperature concentration diagram for polymer solutions [102, 103]. $\tau = (T - \Theta)/\Theta$; ($\Theta$: Θ-temperature, c monomer concentration (see Table 5)

interactions, blobs of Gaussian conformation are formed when approaching the regime I'. The blob size or correlation length ξ scales with τ^{-1}. At the borderline between I and I', ξ becomes equal to R_g, and the whole chain exhibits an unperturbed random walk conformation.

With respect to the scattering behavior at $Q\xi > 1$, the domains of Gaussian conformation and in the opposite case ($Q\xi < 1$) the domains of swollen chain conformation are probed. The crossover between both regimes is expected to occur at $Q^* = 1/\xi$.

In Table 5, some scaling predictions for $\langle R_g^2 \rangle$ are $\langle \xi \rangle^2$ summarized, which meanwhile have been confirmed by elastic small-angle neutron scattering [104, 107–109].

Similar to the case of good solvent conditions, the complete coherent dynamic structure factor $S(Q, t)$ is not available in the transition range of the regimes I and I'. However, the crossover behavior becomes accessible via the initial slope as a function of Q and τ [105]. Typical results of this treatment are shown in Fig. 39, where

$$\Omega_{red}(Q, \tau) \equiv \frac{T}{\Theta} \frac{\eta_s(\Theta)}{\eta_s(T)} \frac{\Omega_z(Q, \tau)}{\Omega_z(Q, 0)} \tag{113}$$

Table 5. Scaling predictions for the mean square radius of gyration $\langle R_g^2 \rangle$ and the mean square correlation lengths $\langle \xi^2 \rangle$ in the different regimes (see Fig. 38) of polymer solutions [102–104]

Regime	$\langle R_g^2 \rangle$	$\langle \xi^2 \rangle$
I'	N	—
I	$N^{6/5}\tau^{2/5}$	τ^{-2}
II	$Nc^{-1/4}\tau^{1/4}$	$c^{-3/2}\tau$
III	N	c^{-2}

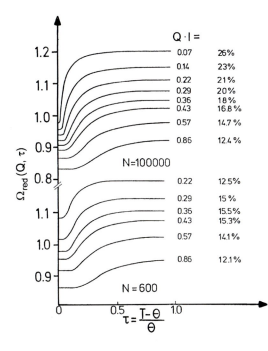

Fig. 39. Crossover from Θ- to good solvent conditions in dilute solutions. Calculated characteristic frequencies, normalized to Θ-conditions, as dependent on reduced temperature for two different chain lengths N at various values of $(Q\ell)$. To the *right* of each curve the increase in $\Omega_{red}(q, \tau)$ between $\tau = 0$ and $\tau = 0.9$ is given

is plotted vs. τ for two different chain lengths and pre-averaged Oseen tensor. This plot shows that:

1. For large numbers of segments N and small values $Q\ell$, a sharp crossover from Θ- to good solvent conditions occurs;
2. For a given chain length, the sharpness of the crossover as well as the increase in $\Omega_{red}(Q, t)$ are diminished with increasing $(Q\ell)$-values;
3. The reduction of N reduces considerably the increase in $\Omega_{red}(Q, \tau)$ and smears out the sharpness of the transition;

4. Effects resulting from the violation of the conditions $N \to \infty$ and $Q\ell \ll 1$, which were used to discuss (see above) the Q-dependence and the prefactors of the characteristic frequencies in the limits of Θ- and good solvent states.

Despite the problems due to non-asymptotic conditions, the relation $\xi(\tau) = 1/Q^* \sim 1/\tau$ is regained from the curves of Fig. 39 with reasonable accuracy.

5.1.2 NSE Results from Dilute Solutions of Linear Polymers

Θ-Conditions

NSE investigations on the segmental diffusion of chain molecules in dilute solutions under Θ-conditions have been performed on three systems [12, 40, 110] (see Table 6). In Fig. 40a, typical spectra obtained from poly(dimethyl-siloxane) (PDMS) in d-bromobenzene at $T = 357$ K are shown together with the outcome of fitting the scattering law of the Zimm model (see Table 1) simultaneously to all spectra. If the numerical prefactor in Eq. (84) for $\Omega_Z(Q)$ is used as an adjustable parameter in the fitting procedure, an excellent agreement between the experimental findings and the theoretical predictions has to be stated. Thus, it is not surprising that in a scaling representation, where $S(Q, t)/S(Q, 0)$ is plotted vs. the scaling variable $(\Omega_Z(Q)t)^{2/3}$, the data follow a master curve (see Fig. 40b). If the scattering law of the Zimm model is fitted to the spectra at different Q-values, the characteristic frequencies $\Omega_Z(Q)$ are found to obey the predicted Q^3-behavior (see Fig. 41). A common fit, where the power of Q was introduced as a second adjustable parameter in a "Zimm-like" theoretical scattering law, gives $\Omega(Q) \sim Q^{3.0 \pm 0.1}$. On the other hand, the corresponding asymptotic limit of the scattering law of the Zimm model [Eq. (88)] does not fit the experimental data, although with increasing Q-values a gradual tendency towards the long-time limit is observable [110]. Concerning the line shape and the Q dependence of the characteristic frequencies, a similar good agreement between the experimental spectra and the predictions of the Zimm model are also found for poly(styrene) PS in two different Θ-solvents. Furthermore, in the case of PS/d-cyclohexane the transition from $\Omega(Q) \sim Q^3$ to $\Omega(Q) \sim Q^2$ becomes visible in the Q-range of observation $0.02 \leqslant Q/\text{Å}^{-1} \leqslant 0.14$

Fig. 40a, b. NSE spectra of a dilute solution under θ-conditions (PDMS/d-bromobenzene, $T = \Theta = 357$ K). **a** $S(Q, t)/S(Q, 0)$ vs time t; **b** $S(Q, t)/S(Q, 0)$ as a function of the Zimm scaling variable $(\Omega(Q)t)^{2/3}$. The *solid lines* result from fitting the dynamic structure factor of the Zimm model (s. Table 1) simultaneously to all experimental data using T/η_s as adjustable parameter.

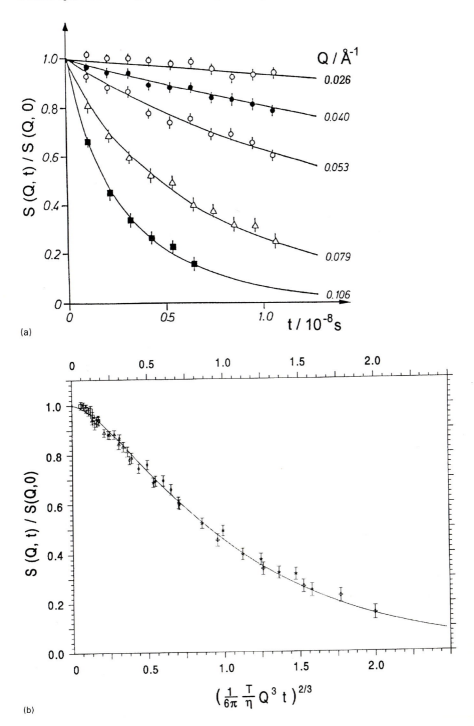

(a)

(b)

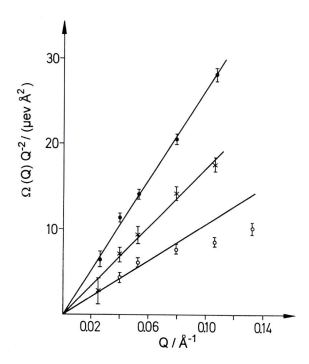

Fig. 41. $\Omega(Q)/Q^2$ vs. Q for different dilute solutions under Θ-conditions ● PDMS/d-bromo-benzene, T = 357 K; × PS/d-methylcyclohexane T = 341 K; ○ PS/d-cyclohexane T = 311 K

Table 6. NSE results from dilute Θ-solutions

	T/K	$\Omega_{exp}(Q)/\Omega_z(Q)$
PDMS/d-bromobenzene	357	0.87
PS/d-methylcyclohexane	341	0.43
PS/d-cyclohexane	311	0.50

at larger Q (see Fig. 41). For all three systems the absolute values of $\Omega_Z(Q)/Q^3$ are smaller than expected from the theory (see Table 6), even if the theoretical $\Omega_Z(Q)$ is calculated on the basis of the pre-averaged Oseen tensor. The discrepancies are much more drastic for PS (~ 50–60%) than for PDMS (14%). The reasons for these deviations are not clear.

Obviously, in the case of PS these discrepancies are more and more reduced if the probed dimensions, characterized by $2\pi/Q$, are enlarged from microscopic to macroscopic scales. Using extremely high molecular masses the internal modes can also be studied by photon correlation spectroscopy [111, 112]. Corresponding measurements show that – at two orders of magnitude smaller Q-values than those tested with NSE – the line shape of the spectra is also well described by the dynamic structure factor of the Zimm model (see Table 1). The characteristic frequencies $\Omega_Z(Q)$ also vary with Q^3. However, their absolute values are only 10–15% below the prediction.

In the case of dynamic mechanical relaxation the Zimm model leads to a specific frequency (ω) dependence of the storage [$G'(\omega)$] and loss [$G''(\omega)$] part of the intrinsic shear modulus [$G^*(\omega)$] [1]. The smallest relaxation rate $1/\tau_Z$ [see Eq. (80)], which determines the position of the $\log[G'(\omega)]$ and $\log[G''(\omega)]$ curves on the logarithmic ω-scale relates to $\Omega_Z(Q)$, if R_e^3/τ_Z is compared with $\Omega(Q)/Q^3$. The experimental results from dilute PDMS and PS solutions under Θ-conditions [113, 114] fit perfectly to the theoretically *predicted* line shape of the components of the modulus. In addition $1/\tau_Z$ is in complete agreement with the theoretical prediction based on the pre-averaged Oseen tensor.

Good Solvent Conditions

In contrast to Θ-conditions a large number of NSE results have been published for polymers in dilute good solvents [16, 110, 115–120]. For this case the theoretical coherent dynamic structure factor of the Zimm model is not available. However, the experimental spectra are quite well described by that derived for Θ-conditions. For example, see Fig. 42a and 42b, where the spectra $S(Q, t)/S(Q, 0)$ for the system PS/d-toluene at 373 K are shown as a function of time t and of the scaling variable $(\Omega_Z(Q)t)^{2/3}$. As in Fig. 40a, the solid lines in Fig. 42a result from a common fit with a single adjustable parameter. No contribution of Rouse dynamics, leading to a dynamic structure factor of combined Rouse-Zimm relaxation (see Table 1), can be detected in the spectra. Obviously, the line shape of the spectra is not influenced by the quality of the solvent. As before, the characteristic frequencies $\Omega(Q)$ follow the Q^3-power law, which is a strong indication for Zimm relaxation. These findings are in complete agreement with the results of dynamic light-scattering studies [111, 121, 122]. In contrast to these scattering results, the line shapes of $\log[G'_R(\omega)]$ and $\log[G''_R(\omega)]$ as a function of $\log \omega\tau_Z$ differ from those observed under Θ-conditions. For good solvents the spectra look more Rouse-like [1]. The absolute values of $\Omega(Q)$ again are considerably smaller than those predicted by the theory. On the basis of the pre-averaged Oseen tensor and the expected increase of $\Omega_Z(Q)$ by a factor 1.34 compared to Θ-conditions, 20–30% are missing in the case of PDMS and 40–50% in the case of PS [123]. Thus, the deviation of the prefactor from the theory seems to depend not only on the specific polymer but also on details of the interaction between polymer and solvent.

Particular attention was placed on the crossover from segmental diffusion to the center of mass diffusion at $Q \simeq 1/R_g$ and to the monomer diffusion at $Q \simeq 1/\ell$, respectively, by Higgins and coworkers [119, 120]. While the transition at small Q is very sharp (see Fig. 43, right side), a broader transition range is observed in the regime of larger Q, where the details of the monomer structure become important (see Fig. 44). The experimental data clearly show that only in the case of PDMS does the range $\Omega(Q) \sim Q^3$ exceed $Q = 0.1 \, \text{Å}^{-1}$, whereas in the case of PS and polytetrahydrofurane (PTHF) it ends at about $Q = 0.06–0.07 \, \text{Å}^{-1}$. Thus, the experimental Q-window to study the internal dynamics of these polymers by NSE is rather limited.

Fig. 42. NSE spectra of a dilute solution under good solvent conditions (PDMS/d-toluene, $T = 373$ K). **a** $S(Q, t)/S(Q, 0)$ vs. time t; **b** $S(Q, t)/S(Q, 0)$ as a function of the Zimm scaling variable $(\Omega(Q)t)^{2/3}$. The *solid lines* result from fitting the dynamic structure factor of the Zimm model (see Table 1) simultaneously to all experimental data using T/η_s as adjustable parameter

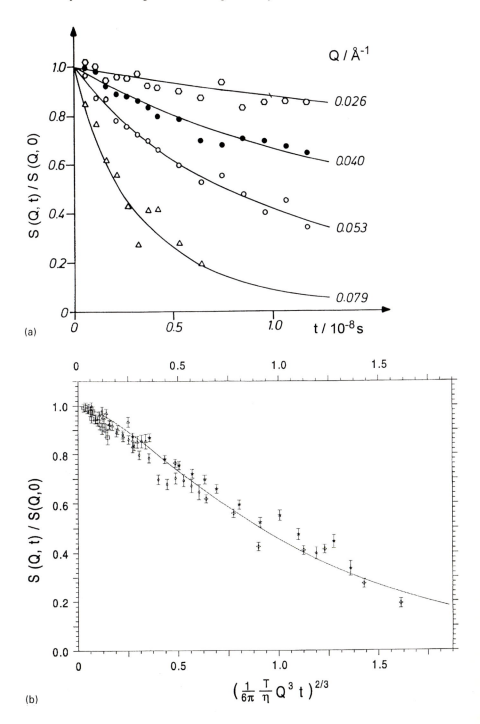

(a)

(b)

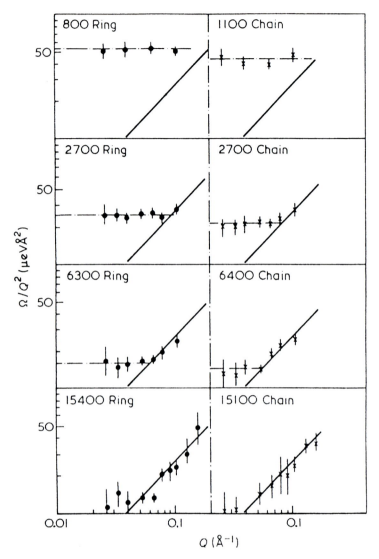

Fig. 43. Experimentally determined behavior of Ω/Q^2 for various samples of linear and cyclic PDMS. The *broken lines* indicates the experimental value of D_Z whilst the *solid lines* show the theoretical Q^3 line as given by Eq. (84) [120]. (Reprinted with permission from [120]. Copyright 1983 Elsevier Science Ltd., Kidlington, UK)

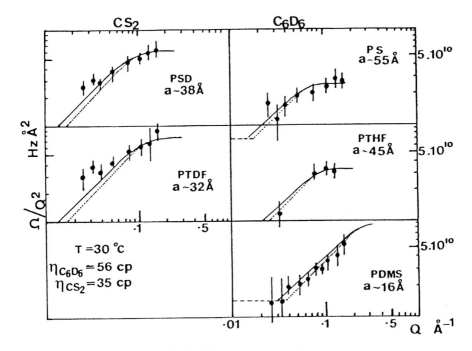

Fig. 44. Double logaritmic plot of $\Omega(Q)/Q^2$ vs. Q for various dilute solutions under good solvent conditions visualize the crossover from segmental to monomer diffusion [119]. The *solid lines* result from fitting the theoretical predictions of Akcasu et al. [94] to the experimental data using B = 0.38 and η_s and a $\equiv \ell$ as adjustable parameters. The *dotted lines* are the corresponding predictions for Θ conditions. (Reprinted with permission from [119]. Copyright 1981 American Chemical Society, Washington)

Crossover from Θ- to Good Solvent Conditions

The crossover from Θ- to good solvent conditions in the internal relaxation of dilute solutions was investigated by NSE on PS/d-cyclohexane (Θ = 311 K) [115] and on PDMS/d-bromobenzene (Θ = 357 K) [110]. In Fig. 45 the characteristic frequencies $\Omega_{red}(Q, \tau)$ (113) are shown as a function of $\tau = (T - \Theta)/\Theta$. The $\Omega_Z(Q, \tau)$ were determined by fitting the theoretical dynamic structure factor S(Q, t)/S(Q, 0) of the Zimm model (see Table 1) to the experimental data. This procedure is justified since the line shape of the calculated coherent dynamic structure factor provides a good description of the measured NSE-spectra under Θ- as well as under good solvent conditions.

In both cases, qualitatively a good agreement with the theoretical predictions (see Fig. 39) has to be stated. The smaller the Q-value, the stronger the percentage increase at fixed τ and the smaller the temperature range up to the maximum increase.

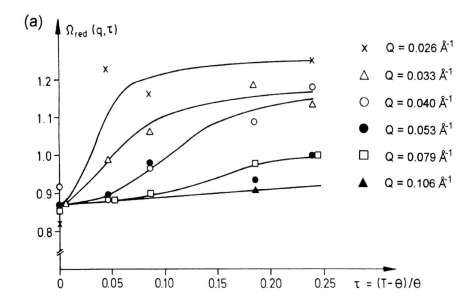

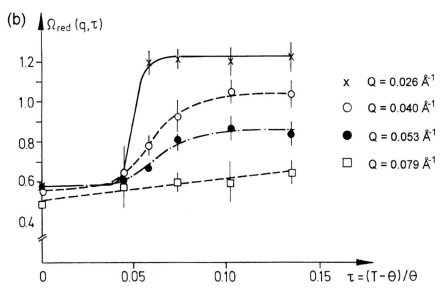

Fig. 45a, b. Segmental diffusion in dilute solutions at the crossover from Θ- to good solvent conditions. Reduced characteristics frequencies $\Omega_{red}(Q, \tau)$ vs. $\tau \equiv (T - \Theta)/\Theta$ at different Q-values **a** PDMS/d-bromobenzene; **b** PS/d-cyclohexane. (**b** reproduced with permission from [115]. Copyright 1980 The American Physical Society, Maryland)

However, quantitative differences between both systems and between experimental results and theoretical predictions also become obvious. For the PDMS system the highest increase of $\Omega_{red}(Q, \tau)$ is 44% and three times larger than expected for a chain length of about 800 monomer units, whereas for the PS system the highest increase amounts to more than 100%. In addition, for a chain length of 600 monomer units the calculations yielded temperature breadth of the crossover extending up to $\tau = 0.5$ for small Q's, whereas experimentally it occurs on a scale compressed by one order of magnitude. A direct comparison between theory and experiment is shown in Fig. 46, where $\Omega_{red}(Q, \tau)$ from the PDMS system is plotted as a function of Q for five τ-values. This plot demonstrates that the discrepancies between experiment and theory mainly occur at smaller Q-values. In other words, on larger length scales the segmental diffusion becomes considerably more enhanced than expected from the theoretical approach.

If the crossover points $Q^*(\tau)$ are determined from Fig. 45, taking the τ-values at half-step height, $Q^*(\tau) = 1/\xi(\tau) = (0.7 \pm 0.2)\tau$ is obtained in the case of the PS system. This has to be compared with static value $Q_s^*(\tau) \cong 1.6\tau$, derived from the same polymer solvent system by elastic neutron scattering [103]. As long as no corresponding data from other polymer solvent systems are available, the final decision as to whether static and dynamic scaling lengths coincide or not, is still open.

The crossover behavior of a dilute high molecular mass PS/cyclohexane solution was also studied by photon correlation spectroscopy probing the regimes of center of mass and segmental diffusion [124]. The temperature dependence of the reduced characteristic frequencies, which was analyzed in terms of a modified blob model for the static equilibrium distribution function [125], is found to be in very good agreement with the theoretical calculations in the low Q-range. However, in the regime of segmental diffusion $(Q\langle R_g^2 \rangle^{1/2} \geqslant 4$ the general trend of the data, which, similar to the NSE data decrease with increasing Q for $T > \Theta$ (compare Fig. 45), is not reproduced by the refined theory, which suggests a plateau behavior with continuously increasing height, even before good solvent conditions are reached.

On macroscopic length scales, as probed for example by dynamic mechanical relaxation experiments, the crossover from Θ- to good solvent conditions in dilute solutions is accompanied by a gradual variation from Zimm to Rouse behavior [1, 126]. As has been pointed out earlier, this effect is completely due to the coil expansion, resulting from the presence of excluded volume interactions.

To summarize the results of NSE investigations on dilute homopolymer solutions, the following conclusions have to be drawn:

1. The line shape of the spectra is independent from the solvent conditions and in complete agreement with the calculations based on the Zimm model
2. For Θ- as well as for good solvent conditions, the line shapes of the spectra follow a master curve when $S(Q, t)/S(Q, 0)$ is plotted vs. $(\Omega(Q)t)^{2/3}$, where $\Omega(Q) \sim Q^3$ is the characteristic frequency of the Zimm relaxation

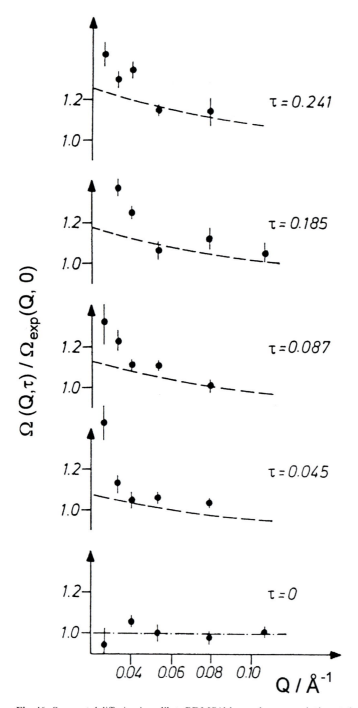

Fig. 46. Segmental diffusion in a dilute PDMS/d-bromobenzene solution at the crossover from Θ to good solvent conditions. Reduced characteristic frequencies Ω_{red} (Q, τ) vs. Q at different τ-values. Comparison between experimental results (●) and theoretical predictions (---). according to [98].

3. In contrast to the theoretical predictions, the absolute values of $\Omega(Q)$, normalized with respect to the solvent viscosity η_s and the temperature T, are non-universal. They vary from polymer to polymer and are always smaller than expected from theory

4. The crossover from Θ- to good solvent conditions leads at constant $\tau = (\tau - \Theta)/\Theta$ to increasing $\Omega(Q,\tau)/Q^3$ with decreasing Q. Qualitatively, this effect is well described in the framework of the blob model using the method of the first cumulant, proposed by Akcasu and coworkers

5. For PS/d-cyclohexane, the only system where comparable static and dynamic data from the crossover regime are available, the static and dynamic crossover points $Q^*(\tau) = 1/\xi(\tau)$ do not coincide

5.2 Dilute Solutions of Cyclic Polymers

Chain and ring macromolecules are topologically distinct. Thus it is not surprising that many differences in their microscopic properties are observed [127]. Besides many other experimental techniques, which were applied to specify these differences, NSE was used to compare the center of mass diffusion and the internal relaxation of linear and cyclic PDMS systems in dilute solutions under good solvent conditions [120, 128, 129]. An important parameter for these investigations was the molecular mass, which was varied between 800 and 15400 g/mol and which was almost identical for the corresponding linear (L) and ring (R) systems.

In Fig. 43 the characteristic frequencies $\dot{\Omega}(Q)$ divided by Q^2 are plotted vs. Q. Qualitatively the behavior of corresponding solutions is very similar. The cross over from center of mass to segmental diffusion is seen for molecular masses of about 6000 g/mol, whereas for 800 g/mol and 15400 g/mol only the center of mass diffusion or the internal relaxation can be observed. The Q-value (Q^*), which separates both regimes, in general is smaller for the linear than for the cyclic polymers, when similar molecular masses are considered. This is due to the fact that for linear and cyclic PDMS containing the same number of monomer units $\langle R_g^2 \rangle_L^{1/2} > \langle R_g^2 \rangle_R^{1/2}$ is valid [128]. The ratio of the center of mass diffusion coefficients D_L/D_R is found to be 0.84 ± 0.02, which compares favorably with 0.85, predicted in the case of absent-free draining and excluded volume effects [130]. The absolute values of D_L and D_R agree quite well with those obtained by boundary spreading [131] and quasi-elastic light-scattering techniques [132].

With respect to the segmental diffusion, the characteristic frequencies of the cyclic systems vary with Q^3 as in the case of the linear chains in dilute solution (see Sect. 5.1.2). The absolute values are independent of the topology of the polymers and their molecular masses, and thus exhibit the same deviations from the theoretical predictions that have just been pointed out for dilute solutions of linear homopolymers.

5.3 Dilute Solutions of Star-Shaped Polymers

5.3.1 Theoretical Outline—Influence of the Static Structure on the Dynamics

Star shaped macromolecules are polymers, where the one end of $f > 2$ (f functionality of the star) linear chains is chemically attached by covalent bonds to a small central linker unit, are the simplest form of branched polymers. Modern anionic polymerization techniques allow us to synthesize star systems with a large number of nearly monodisperse arms [133, 134].

Compared with linear macromolecules, the monomer density profile of such regular multiarm stars is highly non-uniform [135, 136]. Near the center it can approach that of a melt, even in dilute solutions. Progressing outwards from the center towards the edge it is lowered considerably, since more space is available for each of the arms to occupy. Finally, as one approaches the edge of the star, the monomer density will decrease fairly abruptly. On the basis of scaling ideas this may be pictured in terms of the concept of semi-dilute solutions with a blob size decreasing from the rim to the center [137–139]. These structural peculiarities are directly mirrored in the static scattering behavior [140–143].

The dynamics of highly diluted star polymers on the scale of segmental diffusion was first calculated by Zimm and Kilb [143] who presented the spectrum of eigenmodes as it is known for linear homopolymers in dilute solutions [see Eq. (77)]. This spectrum was used to calculate macroscopic transport properties, e.g. the intrinsic viscosity [145]. However, explicit theoretical calculations of the dynamic structure factor [S(Q, t)] are still missing at present. Instead of this the method of first cumulant was applied to analyze the dynamic properties of such diluted star systems on microscopic scales.

The best insight into the relaxation behavior of star polymers in dilute solution can be expected if, in addition to the whole star system, different parts of the star are considered separately. This can be achieved easily by neutron scattering techniques on systems where not only the entity of arms, but also single arms, the core or shell parts are labelled by proton deuterium exchange. With respect to the core-shell labelling it is convenient to build up the arms as diblock copolymers of A–B type with protonated or deuterated but otherwise chemically identical A and B blocks.

Quasielastic Scattering of a Star with Diblock Arms in Solution

If the partially labelled star/solvent system is considered as an incompressible ternary solution, the double differential cross section $\partial^2\sigma/\partial\Omega\partial E$ can be written as

$$\frac{\partial\sigma}{\partial\Omega\partial E} = n \int_{-\infty}^{\infty} dt\, e^{-i\omega t} \sum_{\alpha=1}^{f} \sum_{\beta=1}^{f} \sum_{j=1}^{N_\alpha} \sum_{k=1}^{N_\beta} K_j^\alpha K_k^\beta \left\langle e^{i\underline{Q}(\underline{r}_j^\alpha(t) - \underline{r}_k^\beta(0))} \right\rangle. \tag{108}$$

n is the number of stars per unit volume; N_α, N_β is the number of segments in both blocks A and B; K_j^α, K_k^β is the contrast between the solvent molecule and segment j in block A and segment k in block B, respectively. Following Eq. (6) K_j^α is given by

$$K_j^\alpha = \left(\bar{b}_p^\alpha - b_s \frac{v_p^\alpha}{v_s} \right) \tag{109}$$

where $\bar{b}_p^\alpha, b_s$ represents the scattering lengths of the monomer in block A and the solvent molecule; v_p^α, v_s are the corresponding specific volumina.

Introducing partial intermediate structure factors

$$S_{\alpha\beta}(\underline{Q}, t) = \sum_j \sum_k \left\langle e^{i\underline{Q}(\underline{r}_j^\alpha(t) - \underline{r}_k^\beta(0))} \right\rangle, \tag{110}$$

Eq. (108) may be written as

$$\frac{\partial^2 \sigma}{\partial \Omega \, \partial E} = n \int_{-\infty}^{\infty} dt \, e^{-i\omega t} \sum_\alpha \sum_\beta K_\alpha K_\beta S_{\alpha\beta}(\underline{Q}, t). \tag{111}$$

Time Evolution of the Partial Dynamic Structure Factors

If all memory effects are neglected, in the framework of the linear response theory the equation of motion for $S_{\alpha\beta}(\underline{Q}, t)$ can be written as [146, 147].

$$\frac{\partial}{\partial t} S_{\alpha\beta}(\underline{Q}, t) = - \sum_\gamma \Omega_{\alpha\gamma}(\underline{Q}) S_{\gamma\beta}(\underline{Q}, t). \tag{112}$$

$\Omega_{\alpha\gamma}(\underline{Q})$ is the element $\alpha\gamma$ of the first cumulant or initial decay matrix $\underline{\underline{\Omega}}(\underline{Q})$ defined by

$$\underline{\underline{\Omega}}(\underline{Q}) \equiv \lim_{t \to 0} \frac{d}{dt} \ln \underline{\underline{S}}(Q, t). \tag{113}$$

The system of coupled differential equations (112) is solved by

$$S_{\alpha\beta}(\underline{Q}, t) = a_{\alpha\beta}^1 e^{-\Gamma_1 t} + a_{\alpha\beta}^2 e^{-\Gamma_2 t} \tag{114}$$

with an optic like mode

$$\Gamma_1 = \Omega_{av} + \sqrt{\Omega_{av}^2 - \Delta\Omega}. \tag{115}$$

resulting from motion of both blocks against each other and a diffusive mode

$$\Gamma_2 = \Omega_{av} - \sqrt{\Omega_{av}^2 - \Delta\Omega}. \tag{116}$$

Ω_{av} and $\Delta\Omega$, which both depend on Q, are given by

$$\Omega_{av}(Q) = \tfrac{1}{2}(\Omega_{\alpha\alpha}(Q) + \Omega_{\beta\beta}(Q)) \tag{117}$$

$$\Delta\Omega(Q) = \Omega_{\alpha\alpha}(Q)\Omega_{\beta\beta}(Q) - \Omega_{\alpha\beta}(Q)\Omega_{\beta\alpha}(Q). \tag{118}$$

The spectral weights $a_{\alpha\beta}^i$ are functions of Γ_1 and Γ_2, of the partial static structure factors and of the elements of the $\underline{\underline{\Omega}}(Q)$. On the other hand, $\underline{\underline{\Omega}}$ is related to the generalized mobility matrix $\underline{\underline{\mu}}(Q)$ by [compare Eq. (94)].

$$\underline{\underline{\Omega}}(Q) = k_B T Q^2 \underline{\underline{\mu}}(Q)/\underline{\underline{S}}(Q,0). \tag{119}$$

Thus, the calculation of $\underline{\underline{\Omega}}(Q)$ requires the knowledge of the partial static structure factors $S_{\alpha\beta}(Q,t)$ and the elements $\mu_{\alpha\beta}(Q)$ of the mobility matrix $\underline{\underline{\mu}}(Q)$, which itself depend on $S_{\alpha\beta}(Q,0)$.

Static Structure Factor of Star Polymers

According to Benoit and Hadziioannou [148] the partial structure factors of a f functional star polymer with diblock arms of A–B-type in the Gaussian approximation are given by

$$S_{\alpha\alpha}(Q,0) = f P_\alpha(Q) + f(f-1)A_\alpha^2(Q)$$

$$S_{\beta\beta}(Q,0) = f P_\beta(Q) + f(f-1)A_\beta^2(Q)e^{-2Q^2\langle R_{g,a}^2\rangle}$$

$$S_{\alpha\beta}Q,0) = S_{\beta\alpha}(Q) = f A_\alpha^2(Q) + f(f-1)A_\alpha^2(Q)e^{-2Q^2\langle R_{g,a}^2\rangle} \tag{120}$$

$P_\gamma(Q)$ $(\gamma = \alpha, \beta)$ is the generalized Debye structure factor

$$P_\gamma(Q) = 2N_\gamma^2 \int_0^1 du(1-u)\exp\{-Q^2\langle R_{g,\gamma}^2\rangle u^{2\nu}\} \tag{121}$$

with $\nu = 1/2$ for Θ and $\nu = 2/3$ for good solvent conditions. A_γ $(\gamma = \alpha, \beta)$ is given by

$$A_\gamma(Q) = N_\gamma \int_0^1 du \exp\{-Q^2\langle R_{g,\gamma}^2\rangle u^{2\nu}\}. \tag{122}$$

(121) and (122) are derived from the assumption that the distance distribution between two arbitrary polymer segments j and k is Gaussian independent of the magnitude of the Flory exponent ν.

For a Gaussian star with f uniform arms of N segments one obtains [140]

$$S(Q,0) = \frac{2N^2}{fz^2}\left[(z-1+e^{-z}) + \frac{f-1}{2}(1-e^{-z})^2\right] \tag{123}$$

with $z = Q^2\langle R_{g,a}^2\rangle$, $\langle R_{g,a}^2\rangle$ being the mean square radius of gyration of one arm. The mean square radius of gyration of the whole star is given by

$$\langle R_g^2\rangle = \frac{3f-2}{f}\langle R_{g,a}^2\rangle. \tag{124}$$

Mobility Matrix

The evaluation of the generalized mobility matrix $\underline{\underline{\mu}}(Q)$ follows the lines outlined in Section 5.1.1. The elements $\mu_{\alpha\beta}(Q)$ can be approximated to high accuracy [100] by sums of a preaveraged term

$$\mu_{\alpha\beta}^{pre}(Q) = \frac{1}{3\pi^2 \eta_s} \int_0^\infty S_{\alpha\beta}(Q^2 + u^2)^{1/2} du \qquad (125)$$

and a correction term $\Delta\mu_{\alpha\beta}(Q)$

$$\Delta\mu_{\alpha\beta}(Q) = \frac{Q^2}{30\pi\eta_s} \int_0^\infty \frac{1}{u^2} [S_{\alpha\beta}(0.72Q^2)^{1/2} - S_{\alpha\beta}(0.72Q^2 + u^2)^{1/2}] du \qquad (126)$$

[see Eq. (99) and (100)]:

Equations (125) and (126) explicitly show that in the initial slope approximation the elements of the generalized mobility matrix can be expressed only in terms of integrals over the corresponding partial static structure factor. Both equations are valid as long as one assumes a Gaussian distance distribution of the distances $r_{ij}^{\alpha\beta}$ between the monomers i on arm α and monomers j on arm β.

Illustration of Two Representative Examples

In order to provide some quantitative insight into the relation between structure and dynamics of star-shaped polymers, first some theoretical calculations for a dilute Gaussian f-arm stars are shown in Fig 47. In a scaled Kratky representation ($Q^2 S(Q,0)$ vs. $z = Q\langle R_{g,a}^2\rangle^{1/2}$, Fig. 47a) for $f = 2$ (linear chains) a monotone increase to the plateau behavior at $z > 1$ is to be seen, whereas for $f > 2$ a maximum occurs at $Qz \simeq 1.2$ before the plateau value is reached. With respect to the dynamics, the deviations of the star relaxation from that of a corresponding linear polymer are visualized most clearly if reduced cumulants $\Omega(Q)/Q^3$ are plotted vs. z or Q (Fig. 47b). In this representation the segmental diffusion of linear polymers follows a line parallel to the abscissa at larger Q or $z > 1$ owing to $\Omega(Q)/Q^3 \sim Q^0$ [see Eq. (84)], whereas for smaller Q or $z < 1$ the crossover to the center of mass diffusion ($\Omega(Q)/Q^3 \sim Q^{-1}$) becomes visible. In contrast, the stars exhibit a minimum at Q- or z-values intermediate between the overall translational and the internal diffusion. The depth of the minimum increases with increasing functionality f. Obviously, the positions of the maxima in Fig. 47a and of the minima in Fig. 47b are strongly correlated, since both are located at $Qz \simeq 1.2$.

The second example deals with dilute solutions of 12-arm stars, where the arm are built up by symmetrical diblocks of protonated and deuterated but otherwise identical monomers [149]. Figure 48 displays the reduced relaxation rates $\Gamma_{1,2}/Q^3$. In addition the effective 1/e decay rates of the dynamic structure

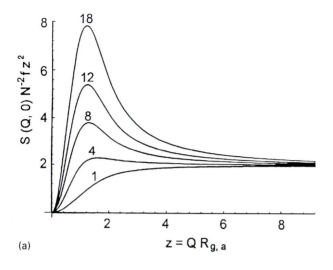

(a)

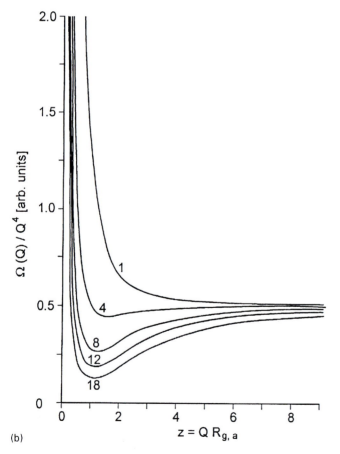

(b)

Fig. 47a, b. Structure and dynamics of star-shaped polymers with different functionalities. **a** Kratky plot of the static structure factor $(S(Q, 0) Q^2$ vs. $Q R_g^a$. **b** $\Omega(Q)/Q^3$ vs. $Q R_g^a$, as derived from Eqs (94) and (123), assuming Rouse dynamics

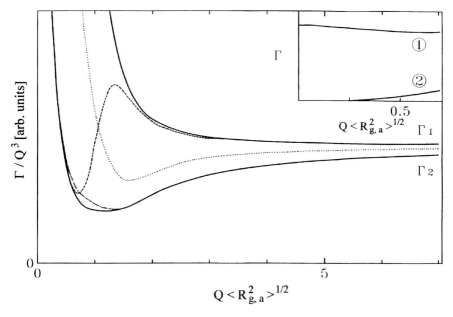

Fig. 48. Reduced relaxation rates $\Gamma_{1,2}/Q^3$ for stars with Gaussian chain conformation. The *insert* represents the rates directly. Note that Γ_1 approaches a constant at lower Q. The *broken lines* give the effective reduced relaxation rates for various contrast conditions. *Dashed line* shell contrast; *dashed-dotted line* core contrast; *dotted line* average contrast. (Reprinted with permission from [154]. Copyright 1990 American Chemical Society, Washington)

factor for three different contrast situations are shown, namely (a) core contrast ($K_1 = 1$, $K_2 = 0$), (b) shell contrast ($K_1 = 0$, $K_2 = 1$), and (c) average contrast ($K_1 = -K_2$). Γ_2/Q^3 exhibits a minimum around $QR_g = 1$ and at low Q describes the translational diffusion of the whole star. This can be seen from the insert of Fig. 48, which displays the non-reduced eigenvalues as a function of QR_g. At low QR_g, Γ_2 increases proportional to Q^2 exhibiting diffuse behavior. On the other hand Γ_1 reaches a finite value at low Q being an optic-like mode describing the relative motion of the star shell with respect to the core. At higher QR_g, the mode Γ_1 does not exhibit the minimum shown by Γ_2. For very large QR_g, both modes converge describing the Zimm relaxation of the arms. On inspecting the results for the dynamic structure factors at different contrast conditions, one finds (see Fig. 49) that for core contrast basically the mode Γ_2 is observed while for shell contrast at higher QR_g the relaxation rate follows Γ_1. For small QR_g a crossover to Γ_2 is observed. Thereby, the reduced relaxation rate passes through a maximum. Finally, for average contrast an intermediate relaxation rate is predicted, which for low QR_g mainly picks up the optic mode.

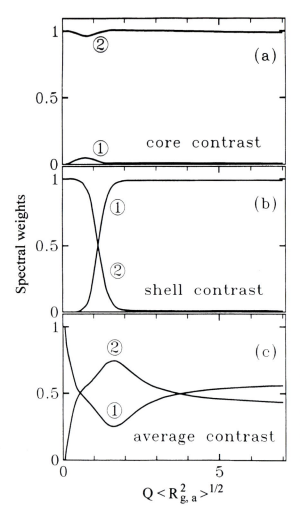

Fig. 49a–c. Spectral contributions of the two modes (①, ②), characterized by Γ_1 and Γ_2, respectively, to the dynamic structure factor for different contrast conditions. (Reprinted with permission from [154]. Copyright 1990 American Chemical Society, Washington)

5.3.2 NSE Results from Stars in Dilute Solution

Completely Labelled Stars

Elastic and quasi-elastic (NSE) neutron scattering experiments were performed on dilute solutions of linear poly(isoprene) (PIP) polymers and of PIP stars (f = 4, 12, 18) [150]. In all cases the protonated polymers were dissolved in d-benzene and measured at T = 323 K, where benzene is a good solvent. Figure 50 shows the results of the static scattering profile in a scaled Kratky representation. In this plot the radii of gyration, obtained from a fit of the experimental data with the corresponding static structure factor [Eq. (123); see Table 7] are used to convert the magnitude of the scattering vector Q to the scaling variable z and to eliminate the intensity scaling factor from the experimental points. The salient feature is the appearance of a maximum in $S(Q)Q^2$, the height of which increases with functionality. It corresponds to an enhancement of the radial segment distribution function G(R) compared with a linear polymer (see inset in Fig. 50). The important Q-range for intermediate- and large-distance correlations (characterized by R_g and f) is covered by all experiments and excellent agreement with the theoretical structure factor is observed. It should further be noted that in the case of the 18-arm stars, the two samples of very different molecular weights produce scattering profiles that fall on one universal line if the Q axis is scaled with the radius of gyration.

A typical example for quasi-elastic measurements on the same system is shown in Fig. 51, where the NSE spectra of an 18-arm star (sample 18IIAA) are reproduced. The solid lines are the line-shape profiles obtained from fitting the individual spectra with the dynamic structure factor of the Zimm model (see Table 1). Obviously, it describes the data very well. The only exceptions are the spectra from the low molecular weight four-arm star, where a crossover to single-exponential decay at small momentum transfers is observed. Figure 52 presents the obtained relaxation rates. They were reduced by Q^3 and are plotted vs. the scaling variable z with use of the experimental radii of gyration for the transformation from Q to z. At large z the reduced rates from all the molecules fall on the same line parallel to the abscissa, an indication of the $\Omega(Q) \sim Q^3$ behavior of the Zimm regime. The absolute values of $\Omega(Q)$ are smaller than predicted by the theory [Eq. (84)]. The deviations are similar as those found for PDMS in good solvent and discussed earlier (see Sect. 5.1.2). The data for the four-arm star stay on this plateau until they bend up at $z \simeq 1.5$. This crossover to $\Omega(Q) \sim Q^2$ behavior coincides with the change of line shape to a single-exponential decay and reflects the onset of translational diffusion. The stars of high functionality exhibit different behavior. They start from the same level at

Table 7. Molecular masses, functionalities, calculated and measured radii of gyration, and translational diffusion coefficients of the investigated polyisoprene stars

Sample	M_w	f	Bonds/arm	$\langle R_g^2 \rangle_{cal}^2$ (Å)	$\langle R_g^2 \rangle_{exp}^{1/2}$ (Å)	D_{exp} (cm²/sec)	D_{theor} (cm²/sec)
18IIAA	22×10^4	18	719	76	96 ± 2		0.6×10^{-6}
18VIIA	6.2×10^4	18	202	40	49 ± 2	$(1.2 \pm 0.2) \times 10^{-6}$	1.2×10^{-6}
12IVA	9.7×10^4	12	475	61	73 ± 1	$(0.9 \pm 0.15) \times 10^{-6}$	0.8×10^{-6}
PIP4	0.5×10^4	4	73	23	42 ± 2	$(1.8 \pm 2) \times 10^{-6}$	1.6×10^{-6}
GQ20	8.4×10^4	1	4940	107			

$\langle R_g^2 \rangle_{cal}^{1/2}$ is the calculated radius of gyration with the parameters for PIP, taken from P.J. Flory (Statistical Mechanics of Chain Molecules, Wiley Intersciences, New York 1969) and the assumption of Gaussian chain statistics.
D_{theor} was calculated according to Ref. [151]

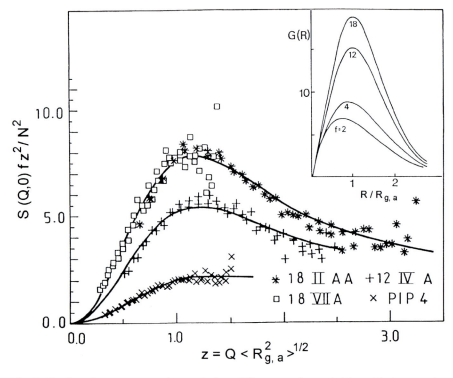

Fig. 50. Small-angle neutron scattering results from different stars in a scaled form. The *lines* are the result of a fit with Eq. (94). *Insert* Related radial segment distribution functions obtained from a Fourier transformation of the theoretical scattering function. (Reprinted with permission from [150]. Copyright 1987 The American Physical Society, Maryland)

large z and the reduced rates pass through a sharp minimum around $z = 1.5$ before they bend upwards into the regime of translational diffusion. Thereby, all data points, also for different f and M_w, appear to follow the same master curve. The width of the minimum (FWHM) amounts to about $\Delta z = 1.5$, its amplitude to 50% of that of the large z-level. From the decay rate in the steep flanks at small Q, the diffusion coefficients of the stars can be derived. As can be seen from Table 7, they compare well with the theoretical prediction by Stockmayer and Fixman [151].

The salient feature of the experimental results is the observation of a pronounced minimum in the $\Omega(Q)/Q^3$ vs. z plot. It occurs at the same position, where the static structure factor in its Kratky representation exhibits its maximum. Furthermore, the reduced line width scales with the scaling variable z in the same way that the static structure factor does. Thus, the occurrence of the minimum is directly related to peculiarities of the star architecture.

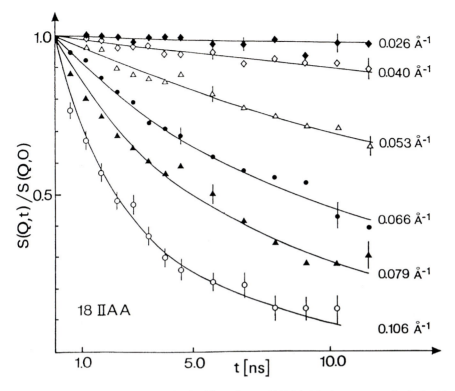

Fig. 51. Neutron spin echo spectra obtained from the star 18 II AA. The *lines* are a result of a fit with the dynamic structure factor calculated by Dubois Violette and de Gennes [34]. (Reprinted with permission from [150]. Copyright 1987 The American Physical Society, Maryland)

In the theory of liquids, such a phenomenon is well known under the term of de Gennes narrowing [152]. There the peak in S(Q) renormalizes the relaxation rate for density fluctuations: $D_{eff} = D/S(Q)$. While in the liquid this is an effect involving different independent particles, here it occurs for the internal density fluctuations of one entity.

A comparison with Burchard's first cumulant calculations shows qualitative agreement, in particular with respect to the position of the minimum. Quantitatively, however, important differences are obvious. Both the sharpness as well as the amplitude of the phenomenon are underestimated. These deviations may originate from an overestimation of the hydrodynamic interaction between segments. Since a star of high f internally compromises a semi-dilute solution, the back-flow field of solvent molecules will be partly screened [40, 117]. Thus, the effects of hydrodynamic interaction, which in general eases the renormalization effects owing to S(Q) [152], are expected to be weaker than assumed in the cumulant calculations and thus the minimum should be more pronounced than calculated. Furthermore, since for Gaussian chains the relaxation rate decreases

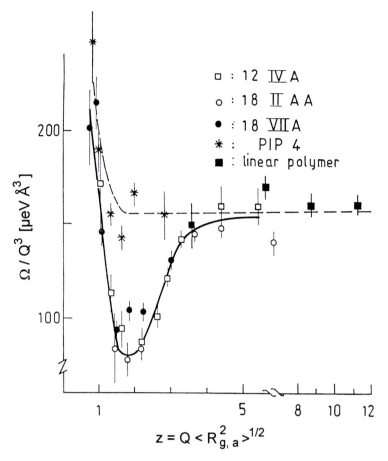

Fig. 52. Reduced relaxation rates Γ/Q^3 as a function of the scaling variable. The *solid line* is a visual aid. (Reprinted with permission from [150]. Copyright 1987 The American Physical Society, Maryland)

compared with swollen chains [115], the screening of the excluded volume interaction – which, owing to the high monomer concentration, will be effective for large distances (stars are only partly swollen) – will further deepen the minimum.

Stars with Single Labelled Arms

In another series of experiments [149] the correlation between structure and dynamics was investigated on dilute solutions of 12-arm PS star systems ($M_w = 14.9 \, 10^4$ g/mol) in d-tetrahydrofurane, where either only one or all 12 arms were protonated (labelled).

As a signature of the star architecture the elastic scattering data of the completely labelled stars exhibit a pronounced peak at $z = 1.5$ in the scaled Kratky representation. In contrast to the PI systems, presented before, the Kratky plot of the measured scattering curve disagrees strongly with the prediction of Eq. (123). The experimental halfwidth of the peak is nearly only 50% of the theoretically predicted one.

This peak structure is completely absent in the scattering profile, obtained from single arm labelled stars (see Fig. 53). In order to demonstrate the asymptotic behavior at large Q's, $S(Q, 0) Q^{1/\nu}$ is plotted vs. Q with $\nu = 3/5$ and $\nu = 2/3$. For smaller chains a plateau is expected for the Flory exponent $\nu = 3/5$. The observed plateauing at higher ν indicates strongly stretched arms. These findings are in good agreement with molecular dynamics simulations on star molecules [135], which also find stretched arm conformations with a Q scaling of the static structure factor of the order of $Q^{-3/2}$.

The reduced characteristic frequencies $\Omega(Q)/Q^3$, as obtained from NSE measurements, are plotted vs. Q for both systems in Fig. 54. The minimum, which occurs for the full stars, does not occur in the system, where the

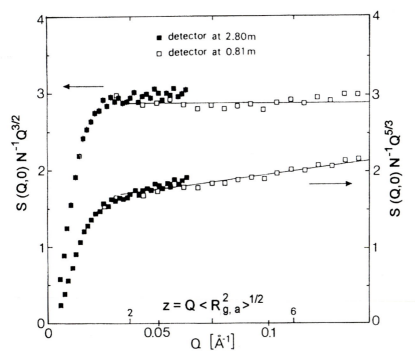

Fig. 53. Small-angle neutron scattering data from the 12-arm polystyrene star PS120A ($M_w = 1.49 \times 10^5$) where the 11 deuterated arms were matched by the solvent THF. In order to demonstrate the asymptotic Q behavior, the data are plotted in a generalized Kratky representation (Iq^α vs. Q with $\alpha = 1.5$ and $5/3$). The *solid line* marks the high Q-plateau. (Reprinted with permission from [149]. Copyright 1989 American Chemical Society, Washington)

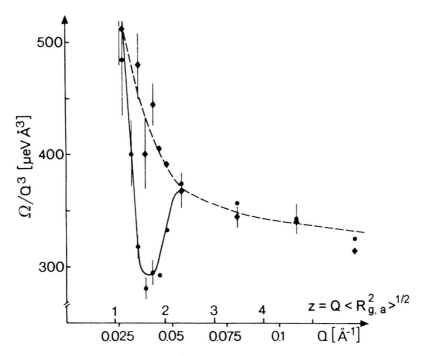

Fig. 54. Reduced relaxation rates Ω/Q^3 for the fully protonated (●) and the one-arm labelled star (◆) as a function of Q. The *lines* are visual aids. (Reprinted with permission from [149]. Copyright 1989 American Chemical Society, Washington)

relaxation of only a single arm is probed. This observation confirms that the minimum is due to the collective motion of the arms. In the case of PS stars the minimum appears to be much narrower ($\Delta z = 0.6$) than that observed for the PIP stars ($\Delta z = 1.5$, see Fig. 52). This effect is as little understood as the different widths of the peaks in the Kratky plot of the corresponding elastic structure factors. However, the discrepancies between the PIP and PS stars do not query, but rather corroborate the principle that there is a close inherent correlation between microscopic structural and dynamic properties of polymers, which is sufficiently described by the ideas underlying Eq. (119).

For the single labelled arm the limiting behavior of $\Omega(Q)/Q^3$ at small and large Q ($z \ll 1$ and $z \gg 1$) agrees with that of the full star. Thereby, the crossover of Ω from Q^3- to the Q^2-dependence is much more gentle than that observed for unattached linear PDMS chains [128] or in the case of the 4-arm PIP stars [150] (see Fig. 52).

Stars with Labelled Cores or Shells

The investigations on partially labelled 12-arm stars in dilute solutions also included PIP systems, where the arms ($M_w = 8000$ g/mol) were built by

Table 8. Details of different labelled 12-arm PIP stars (samples 1, 2, 4 arms of symmetric diblock of protonated and deuterated PI, respectively) [1] barn $\equiv 10^{-24}$ cm^2 [2] evaluated on the basis of the structural model [3] corrected for the ratio of the viscosities of octane and toluene

Sample	1	2	3	4
Systems	HD-star	DH-star	HH-star	DH-star
M_w/g mol^{-1}	9.6 10^4	9.6 10^4	9.6 10^4	9.6 10^4
Solvent	D-octane	D-octane	D-octane	Toluene 90.4% $C_7D_5H_3$ 9.6% $C_7D_3H_5$
Labelling	Core	Shell	Complete	Average
Contrast in barn[1]				
K_1^2	50.1	0.92	50.1	15.5
K_2^2	0.92	50.1		17.2
K_1K_2	-6.83	-6.83		-16.3
T/η_s in K/cP[2]	324 $\pm$ 6	474 $\pm$ 13	389 $\pm$ 7	280 $\pm$ 4[3]

symmetric diblocks of protonated or deuterated, but otherwise identical monomers [154]. As a good solvent d-octane was chosen (see Table 8). Thus, core labelling (H-D-star, sample 1) was present if the protonated blocks were attached to the center and shell labelling (D-H-star, sample 2) in the reversed case. For comparison, a completely labelled star system (H-H-star, sample 3) was also investigated. The average contrast conditions (D-H-star, sample 4) were achieved by replacing the d-octane by a mixture of partially deuterated toluenes (90.4% C_7 D_5 H_3, 9.6% C_7 D_3 H_5). With respect to the total and partial structure factors, scattering from samples 1–3 mainly provides information on $S_{11}(Q,t)$, $S_{22}(Q,t)$ and $S(Q,t)$, respectively, whereas the scattering from sample 4 is sensitive to the interference term $S_{12}(Q,t)$. In Fig. 55 the elastic neutron scattering data for the core, shell and average contrast conditions are shown in a generalized Kratky representation $S(Q,0) Q^{1/\nu}$ vs. Q with $\nu = 3/5$. In all cases the fitted theoretical partial structure factors [see Eq. (120)] for ideal Gaussian ($\nu = 1/2$) and expanded stars ($\nu > 1/2$) are included.

With the core contrast (Fig. 55a) a strong peak is observed as in the case of completely labelled stars [148], while with the average contrast (Fig. 55c) the peak is less intensive. With the shell contrast (Fig. 55b) the scattering pattern is quite different. Its most important feature is the intermediate minimum, which is observed at the same Q-value, where the maxima are found for the core and average contrast conditions.

Similarly to the fully labelled stars [150], the peak structure in the generalized Kratky plot for core and average contrast conditions is also quite well reproduced by calculating S(Q) on the basis of an ideal ($\nu = 1/2$) star conformation. The swollen conformation calculations predict a more diffuse peak in both cases.

At higher Q, however, where the static structure factor reveals the asymptotic power law behavior $S(Q,0) \sim Q^{-1/\nu}$, the assumption of ideal conformation clearly fails. In particular, this is evident for the core (sample 1) and shell contrast conditions (sample 2).

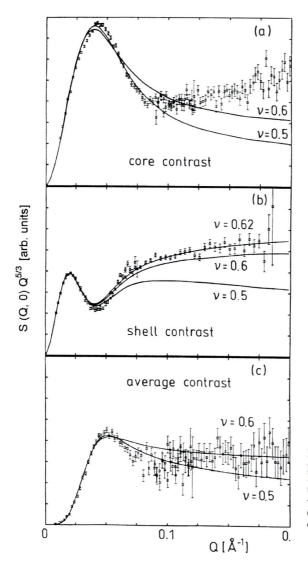

Fig. 55a–c. Generalized Kratky plot $S(Q)\,Q^{5/3}$ vs. Q for diblock PIP stars under different contrast conditions. (Reprinted with permission from [154]. Copyright 1990 American Chemical Society, Washington)

In the case of core contrast, a power law fit of $S(Q,0)$ in the regime $Q > 0.1\,\overset{\circ}{\text{A}}{}^{-1}$ reveals $S(Q,0) \sim Q^{1.4}$, corresponding to a Flory exponent $\nu = 0.71$. An exponent of such magnitude is indicative for a highly stretched chain conformation. For the shell contrast the Flory exponent, for good solvent conditions of linear chains $\nu = 0.6$ provides a reasonably good description of the whole experimental scattering profile. Nevertheless, the agreement is still considerably improved, if ν is slightly increased to 0.62 (see Fig. 55b), indicating a somewhat stronger chain expansion than expected from excluded volume interactions of non-attached linear chains. If these results are compared with that from the single arm, where $v = 0.67$ is found [147] (see above), one realizes that this v-value is in a good approximation the arithmetic mean of the core and the shell contribution.

There is no doubt that the asymptotic large Q-behavior at average contrast conditions agrees better with the expanded chain conformation than with the ideal one. However, since under these contrast conditions the scattering is weak, the data at large Q are affected by large errors, thereby limiting the precision of this statement.

In Fig. 56 the reduced relaxation rates $\Omega(Q)/Q^3$, as obtained from fitting the initial decay of the experimental NSE spectra with the dynamic structure factors of the Zimm model (see Table 1), are shown for the core, shell and average contrast conditions. The dashed and solid lines represent the theoretical predictions using either the partial structure factors of Gaussian stars ($v = 1/2$) with the radii of gyration as determined experimentally by small angle neutron scattering (SANS) (Gaussian model) or the partial structure factors, which were experimentally observed (structural model). In both cases the components of the mobility matrix $\mu_{\alpha\beta}(Q)$ [see Eq. (125) and (126)] were calculated accordingly. The only adjustable parameter was T/n_s, which determines the time scale.

For each contrast condition the Gaussian model (dashed lines) qualitatively displays the general features of the experimental results, but the agreement with the experimental relaxation rates is poor. Overall, the predictions of the Gaussian model exhibit less pronounced structure than actually observed, e.g. for shell and core contrast in Fig. 56, one observes that both the peak and the minimum in the reduced relaxation rates are predicted at the correct position in Q. However, both of them are much weaker than those measured experimentally. Similar observations hold also for the cases of average (Fig. 56c) and full contrast. Compared with the Gaussian model, the structural model provides a superior picture. The agreement with respect to the Q dependence of the characteristic frequencies is nearly quantitative. In particular, both the minimum as well as the maximum of Ω/Q^3 for core and shell relaxation are very well reproduced (Fig. 56). Thus, the relaxation data are well described in terms of the two-mode picture of the RPA model: for Q values above the maximum, the shell relaxation mode may be identified with the "optic" relaxation mode explained earlier. Toward the lower Q, the "acoustic" diffuse mode gains spectral weight, and one observes the down swing of Ω/Q^3 at low Q. On the other hand, the core relaxation behavior reflects nearly exclusively the low-lying second acoustic-like star relaxation mode, which undergoes a minimum where the shell relaxation arrives at the maximum. For average contrast conditions the RPA model predicts a strong low-Q increase of the reduced relaxation rate, since there the optical mode Γ_1 gains weight. This behavior is seen in Fig. 56c, where below $Q = 0.05 \, \text{Å}^{-1}$ the data display the predicted behavior. Since at average contrast the scattering signal approaches zero at $Q = 0$, this mode could not be followed toward the lowest Q values.

While it is evident that the Q dependence of the short-time star relaxation can be exclusively explained on the basis of the star structure, the time scale of these relaxations does not fit into this simple picture. In the model of hydrodynamic interaction this time scale is solely determined by the temperature

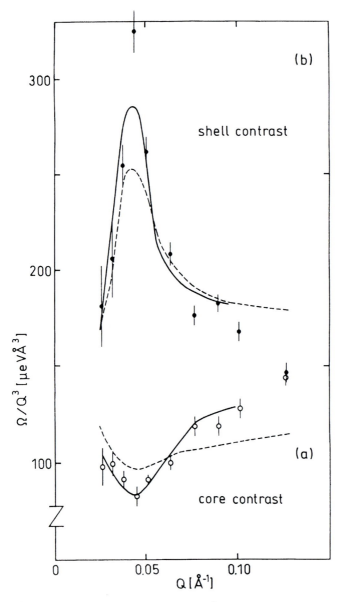

Fig. 56a–c. Reduced relaxation rates Ω (Q)/Q^3 for diblock PIP stars under various conditions and theoretical predictions using the partial structure factors of Gaussian stars (*dashed lines*) or the experimental determined ones (*solid line*). **a** core contrast, **b** shell contrast, **c** average contrast. (Reprinted with permission from [154]. Copyright 1990 American Chemical Society, Washington)

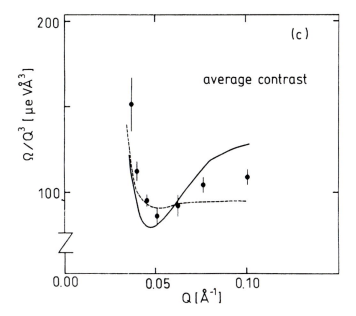

Fig. 56. Continued

T divided by the viscosity of the solvent η_s. For n-octane this number is 837 K/cP at T = 323 K. The results of the fitting process are all below this theoretical value. This is not surprising, since even in the case of dilute solutions of unattached linear chains, the theoretical values are never reached (see Sect. 5.1.2). In addition the experimental T/η_s values differ considerably for the different labelling conditions and the different partial structure factors. Nevertheless, it is interesting to note that T/η_s for the fully labelled stars is within experimental error the arithmetic mean of the corresponding core and shell values.

The theoretical approach described before dealt with the short-time dynamic response of the star molecules. However, in the case of completely labelled stars [148] it was found that the line shape of the Zimm model provides a good description of the NSE spectra not only in the short-time regime (t < 5 ns), but also on longer time scales.

This is different for the star core. Figure 57 provides a comparison of the spectra at two Q-values with those from an equivalent full star (sample 3). Over short periods of time, both sets of spectra nearly coincide. However, over longer periods of time, the relaxation of the star core is strongly retarded and seems to reach a plateau level. This effect may be explained by the occurrence of interarm entanglements as recently proposed by scaling arguments [135].

When the various results obtained by combined elastic and quasielastic neutron scattering measurements on star shaped polymers in dilute solutions

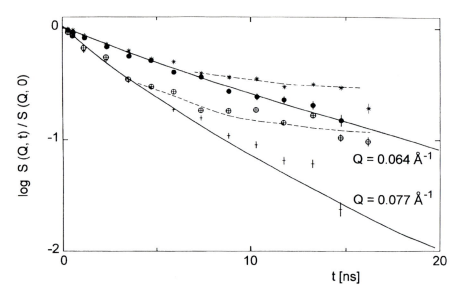

Fig. 57. Relaxation spectra of the fully labelled star (•, +) and the star core (*, ⊕) at two different Q-values. The *solid lines* represent the result of a fit for the Zimm dynamic structure factor to the initial relaxation of the fully labelled star. The *dashed lines* are visual aids showing the retardation of the relaxation for the star core. (Reprinted with permission from [154]. Copyright 1990 American Chemical Society, Washington)

are summarized, the following conclusions arise:

1. With respect to the dynamics of such systems, there is a strong correlation between the structure factor and the initial decay of relaxation on microscopic length scales
2. This correlation can be quantitatively accounted for on the basis of the linear response theory and the RPA
3. Particularly distinct effects occur for $Q\langle R_{g,a}^2\rangle^{1/2} \simeq 1.4$ where $\langle R_{g,a}^2\rangle$ is the mean square radius of gyration of one arm of the star
4. At this position, the reduced relaxation rate $\Omega(Q)/Q^3$ shows a minimum for stars, where the arms are labelled completely or to 50% at the core site. For these systems the Kratky plot of the static structure exhibits a maximum at the same Q-value
5. For single arms neither the minimum in the dynamics nor the maximum in the static response are observed
6. For the star shell the situation is inversed compared with the core (maximum in the dynamic and minimum in the static response)
7. The absolute values of the reduced relaxation rates at $Q\langle R_{g,a}^2\rangle^{1/2} > 1$ are smaller at the core than at the shell of the stars. The corresponding values for the full stars are the arithmetic mean of the core and the shell value

8. The NSE spectra from full stars are well described by the scattering law of the Zimm model in the whole experimental time window (20 ns)
9. Comparing the NSE spectra of complete and core stars, agreement is only found over short periods of time (t < 5 ns). For longer times the spectra from the core tend toward a plateau, indicating interarm entanglement effects

5.4 Semi-dilute Solutions of Linear Homopolymers

5.4.1 Theoretical Outline – Collective Diffusion and Screening of Hydrodynamic Interactions

Transition from Single to Many Chain Behavior

When the monomer concentration exceeds c* [see Eq. (112) and Fig. 38], the different polymer molecules are no longer separated but interpenetrate each other forming a transient network of lifetime τ_g. At constant temperature this network structure is characterized by a concentration-dependent correlation length $\xi(c)$, which may be considered as the mean mesh size of the pseudo gel.

Since the transition from dilute to semi-dilute solutions exhibits the features of a second-order phase transition, the characteristic properties of the single- chain statics and dynamics observed in dilute solutions on all intra-molecular length scales, are expected to be valid in semi-dilute solutions on length scales $r < \xi(c)$, whereas for $r > \xi(c)$ the collective properties should prevail [90].

However, this is true only for good solvent conditions, where $\xi(c)$ is also the correlation length, beyond which both the excluded volume and the hy-drodynamic interaction are screened and self-entanglements (intramolecular knots) are rare.

Under Θ-conditions the situation is more complex. On one side the excluded volume interactions are canceled and $\xi(c)$ is only related to the screening length of the hydrodynamic interactions. In addition, there is a finite probability for the occurrence of self-entanglements which are separated by the average distance $\xi_i(c) = (\xi(c)\ell)^{1/2}$. As a consequence the single chain dynamics as typical for dilute Θ-conditions will be restricted to length scales $r < \xi_i(c)$ [155, 156].

The theoretical description of the dynamic behavior of semi-dilute solutions in the hydrodynamic ($QR_g \ll 1$) and in the many-chain ($Q\xi(c) \ll 1$) regime is based on the two fluid model with the solvent and the solute as independent constituents. The motion of the segments is ruled by the balance between rubber elastic and osmotic restoring forces on the one side and viscous forces on the other. The latter are due to the friction between the polymer molecules and the solvent. The corresponding normalized coherent dynamic structure factor was first derived by Brochard and de Gennes [11, 155, 156] in a way analogous to the approach [157] used earlier for permanent gels. Slightly modified derivations, presented later [158, 159] led to the same result. According to these treatments $S(Q, t)/S(Q, 0)$ is given by

$$\frac{S(Q, t)}{S(Q, 0)} = A_s e^{-\Omega_s(Q)t} + (1 - A_s)e^{-\Omega_l(Q)t} \qquad (127)$$

where the indices s and ℓ refer to the fast (solid like) and slow (liquid like) modes, respectively. Under good solvent conditions for $t < \tau_g$ and $Q\xi(c) < 1$, a "pure" gel like mode occurs with $A_s = 1$ and the relaxation rate

$$\Omega_s(Q) = D_g Q^2 = \frac{k_B T}{6\pi\eta_s\xi(c)} Q^2 \sim Q^2 c^{3/4}. \qquad (128)$$

Since the elastic modulus E_g of the pseudogel is identical to the osmotic modulus E_0, $\Omega_s(Q)$ is equal to the relaxation rate $\Omega_l(Q) = D_c Q^2$ of the hydrodynamic mode ($A_s = 0$), which is relevant for $t > \tau_g$ and $Q\xi(c) \ll 1$.

In contrast, for $Q\xi(c) > 1$ the pseudo gel structure is not yet developed and the segmental dynamics follow the predictions of the Zimm model. $S(Q), t)/S(Q, 0)$ is a universal function of $(\Omega(Q)t)^{2/3}$ with $(\Omega(Q) = 1.3 \Omega_Z(Q)$, just as in dilute solutions on the whole intramolecular length scale [99].

Thus, with decreasing Q a crossover from the single-chain to the collective many-chain relaxation occurs at $Q = 1/\xi(c)$ (see Fig. 58a).

Under Θ-conditions E_g is in general much larger than E_0 (the case $E_g \simeq E_0$ is discussed in Ref. [158]) and for $Q\xi(c) < 1$ the relaxation rates of the hydrodynamic ($t > \tau_g$) and the gel like ($t < \tau_g$) are no longer identical. Whereas the first one is given by $A_s = 0$ and

$$\Omega_l(Q) = D_c Q^2 = \frac{k_B T}{6\pi\eta_s\xi(c)} Q^2 \sim Q^2 c^{1.0}, \qquad (129)$$

the latter is now composed of two contributions with the rates

$$\Omega_s(Q) = D_g Q^2 = \frac{k_B T}{6\pi\eta_s\ell} Q^2 \qquad (130)$$

$$\Omega_l(Q) = D_c t/(\tau_g D_g) \qquad (131)$$

and the amplitude

$$A_s = D_c/D_g = \ell/\xi(c).$$

In the present case $\Omega_l(Q)$ is small compared with 1.

Owing to the appearance of self-entanglements, the regime of single chain relaxation ($Q\xi(c) > 1$) is split into two subregimes. On length scales $1/\xi(c) \leqslant Q \leqslant 1/\xi_i(c)$ (single chain microgel mode) $S(Q, t)$ is similar to that in the case of

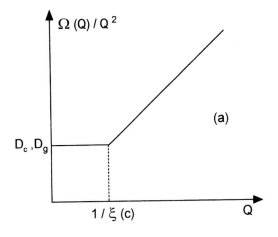

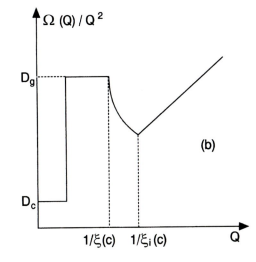

Fig. 58a, b. Segmental diffusion in semi-dilute polymer solutions. Schematic view of the Q-dependence of the relaxation rates $\Omega(Q)$ at a fixed concentration. **a** Good solvent conditions; **b** Θ-conditions. (Reprinted with permission from [168]. Copyright 1994 Elsevier Science B.V., Amsterdam)

the interchain gel-like mode. $\Omega_l(Q)$ remains unchanged; $\Omega_s(Q)$ and A_s are modified to

$$\Omega_s(Q) = D_g^i Q^2 = D_g/(\xi(c)Q) \sim Qc^{-1} \tag{132}$$

and $A_s = D_c/D_g^i$.

Finally for $Q\xi_i(c) \geqslant 1$ the unperturbed (not self-entangled) single-chain relaxation, just known from good solvent conditions, takes place. $S(Q,t)/S(Q,0)$ is a universal function of $(\Omega(Q,t)^{2/3}$ with $\Omega(Q) = \Omega_Z(Q)$. In Fig 58b a schematic plot of the crossover behavior of the segmental dynamics under Θ-conditions is shown.

Transition in the Single Chain Dynamics owing to the Screening
of Hydrodynamic Interactions

It is generally accepted that in semi-dilute solutions under good solvent conditions both the excluded volume interactions and the hydrodynamic interactions are screened owing to the presence of other chains [4, 5, 103]. With respect to the correlation lengths $\xi(c)$ and $\xi_H(c)$ there is no consensus as to whether these quantities have to be equal [11] or in general would be different [160].

The hydrodynamic interaction is introduced in the Zimm model as a pure intrachain effect. The molecular treatment of its screening owing to presence of other chains requires the solution of a complicated many-body problem [11, 160–164]. In some cases, this problem can be overcome by a phenomenological approach [40, 117], based on the Zimm model and on the additional assumption that the average hydrodynamic interaction in semi-dilute solutions is still of the same form as in the dilute case.

If we assume complete screening on length scales larger than $\xi_H(c)$, in this regime the segmental dynamics of a labelled chain in the pseudo gel are expected to follow the predictions of the Rouse model. On the other hand, on length scales smaller than $\xi_H(c)$ the Zimm relaxation should remain unchanged compared with corresponding dilute solutions. With increasing concentration the regime of Rouse relaxation, occurring first at small Q-values, will be extended to larger and larger momentum transfers.

NSE results [40, 117], which will be presented in the next section, showed that the situation is more complex. With increasing concentration, instead of the expected single crossover from Zimm to Rouse, indications for an intermediate Zimm regime were observed.

A model that can take these findings into account is based on the idea that the screening of hydrodynamic interactions is incomplete and that a residual part is still active on distances $r > \xi_H(c)$ [40, 117]. As a consequence the solvent viscosity η_S in the Oseen tensor is replaced by an effective viscosity $\eta_{eff}(r, c)$.

$$\eta_{eff}^{-1}(r, c) = (\eta_s^{-1} - \eta_H^{-1}(c)) \cdot \exp\{ - r/\xi_H(c)\} + \eta_H^{-1}(c). \tag{133}$$

When Eq. (133) is incorporated in the Langevin equation (72) three different dynamic regimes (see Fig. 59) can be distinguished (see Eq. 23).

(1) $Q\xi_H(c) \gg 1$

$$\Omega(Q) \equiv \Omega_z(Q) = \frac{1}{6\pi} \frac{k_B T}{\eta_s} Q^3, \quad \beta = 2/3 \tag{134}$$

unscreened Zimm relaxation (as in the dilute solution),

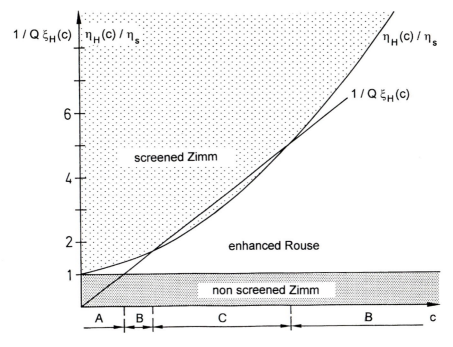

Fig. 59. Incomplete screening of hydrodynamic interactions in semi-dilute polymer solutions. Presentation of different regimes which are passed with increasing concentration. *A,C* Unscreened and screened Zimm relaxation, respectively, *B* enhanced Rouse relaxation. (Reprinted with permission from [12]. Copyright 1987 Vieweg and Sohn Verlagsgemeinschaft, Wiesbaden)

(2) $Q\xi_H(c) \ll 1$ and $Q\xi_H(c) \gg \eta_s/\eta_H(c)$

$$\Omega(Q) = \Omega_z(Q)\left(1 + \frac{\eta_s}{\eta_H(c)} Q\xi_H(c)\right) \quad \beta = 1/2 \tag{135}$$

enhanced Rouse relaxation,

(3) $Q\xi_H(c) \ll 1$ and $Q\xi_H(c) \ll \eta_s/\eta_H(c)$

$$\Omega(Q) = \Omega_z(Q)\frac{\eta_s}{\eta_H(c)} \quad \beta = 2/3 \tag{136}$$

screened Zimm relaxation.

If a residual hydrodynamic interaction over large distances does not exist ($1/\eta_H(c) \equiv 0$), the regime of screened Zimm relaxation vanishes, and only the crossover from unscreened Zimm to enhanced Rouse relaxation remains.

5.4.2 NSE Results from Semi-Dilute Solutions of Linear Homopolymers

Transition from Single Chain to Collective Dynamics

Under good solvent conditions the dynamics of semi-dilute solutions was investigated by NSE using a PDMS/d-benzene system at $T = 343$ K and various concentrations $0.02 \leqslant c \leqslant 0.25$. The critical concentration c^* as defined by (112) is 0.055.

The transition from single- to many-chain behavior already becomes obvious qualitatively from a line shape analysis of the NSE spectra (see Fig. 60) [116]. For dilute solutions ($c = 0.05$) the line shape parameter β is equal to about 0.7 for all Q-values, which is characteristic of the Zimm relaxation. In contrast, in semi-dilute solutions (e.g. $c = 0.18$), β-values of 0.7 are only found at larger Q-values, whereas β-values of about 1.0, as predicted for collective diffusion [see Eq. (128)] are obtained at small Q-values. A similar observation was reported by [163].

Figure 61 presents the $\Omega(Q)/Q^2$ relaxation rates, obtained from a fit with the dynamic structure factor of the Zimm model, as a function of Q. For both dilute solutions ($c = 0.02$ and $c = 0.05$) $\Omega(Q) \sim Q^3$ is found in the whole Q-range of the experiment. With increasing concentrations a transition from Q^3 to

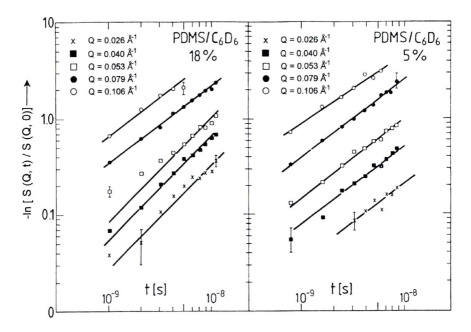

Fig. 60. Crossover from single-chain to many-chain relaxation at $T = 343$ K. Lineshape analysis for PDMS/d-benzene at $c = 5$ and 18%; double logarithmic plot of $-\ln/S(Q, t)/S(Q, 0)$ vs. t/s. (Reprinted with permission from [116]. Copyright 1982 J. Wiley and Sons, Inc., New York)

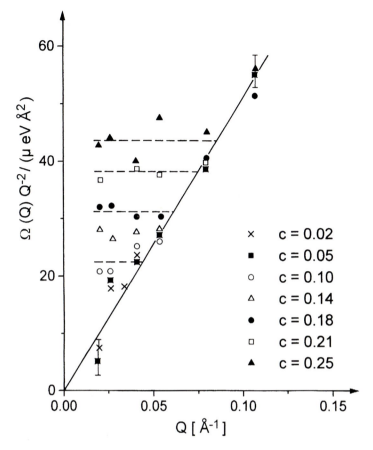

Fig. 61. Crossover from single-chain to many-chain behavior in PDMS/d-benzene at T = 343 K. Characteristic frequencies $\Omega(Q)$ divided by Q^2 as a function of Q. (Reprinted with permission from [116]. Copyright J. Wiley 1982 and Sons, Inc., New York)

Q^2 behavior takes place at decreasing Q. The position of the crossover point $Q^*(c)$ is a direct measure of the dynamic correlation length $\xi(c) = 1/Q^*(c)$. The plateau value at low Q determines the collective diffusion coefficient D_c. A simultaneous fit of all low Q spectra where a simple exponential decay was found led to the concentration dependence and the numerical values of D_c

$$D_c/cm^2\,s^{-1} = (1.3 \pm 0.1)10^{-5} \cdot c^{0.67 \pm 0.05}. \tag{137}$$

The result is shown in Fig. 62 where the different D_c values, obtained from separate fits at each concentration, are also presented. The dynamic correlation length $\xi(c)/\text{Å} = (3.4 \pm 0.7) \cdot c^{-0.67 \pm 0.05}$ is of the same order of magnitude as the value obtained from a static experiment on the PS/d-cyclohexane system [104]. The exponents for the concentration dependence of D_c and $\xi(c)$ are in agreement

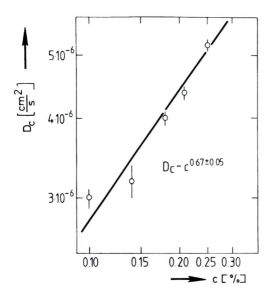

Fig. 62. Collective diffusion coefficient D_C in semi-dilute PDMS/d-benzene systems at $T = 343$ K as a function of the monomer concentration c. (Reprinted with permission from [116]. Copyright 1982 J. Wiley and Sons, Inc., New York)

with the results of dynamic light-scattering experiments [121, 165] probing a much larger length scale. The powers are both smaller than theoretically predicted and found in static neutron scattering experiments. According to [166, 167] this observation does not reflect a fundamental difference between static and dynamic scaling laws, but may result from the finite chain length, which causes the dynamic properties to converge much more slowly to their asymptotic values.

In Fig. 63 the results from PDMS/d-bromobenzene at 357 K (Θ-system) [168] are plotted in the same representation as that used in Fig. 58. In contrast to the good solvent system, the deviations from the unentangled single-chain behavior ($\Omega(Q)/Q^2 \sim Q$) do not lead to Q-independent plateaus but exhibit an increase with decreasing Q-values. According to the arguments of the theoretical section, this behavior is characteristic of the existence of a single chain mode, where self-entanglements have to be taken into account.

The experimental data also show that the crossover to the many-chain regime does not appear in the Q-window accessible to the method. Owing to this lack, the direct experimental evidence that the upturn has to be assigned to a single chain mode and does not result from a collective mode, is still missing. Nevertheless, assuming $D_c \sim c$, and expressing D_g^i by $D_c/(Q\ell)$, from a simultaneous fit $\xi(c)/\mathring{A} = (6.0 \pm 0.5)c^{-1}$ and $\ell/\mathring{A} = 7.4 \pm 0.6$ are derived.

Screening of Hydrodynamic Interactions

The observation of single-chain dynamics in semi-dilute solutions requires contrast matching and labelling. In the case of PDMS this can be achieved using

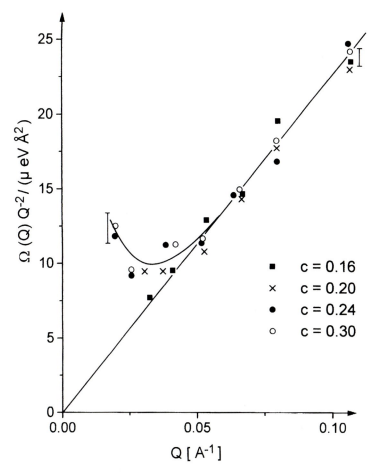

Fig. 63. PDMS/d-bromobenzene at 357 K (Θ-solvent system). Reduced relaxation rate $\Omega(Q)$ of the segmental dynamics as dependent on Q for various concentrations. (Reprinted with permission from [168]. Copyright 1994 Elsevier Science B.V., Amsterdam)

a mixture of 95% deuterated and 5% protonated polymers, dissolved in d-chlorobenzene. Since the scattering length densities of d-PDMS and d-chlorobenzene are almost identical, for all monomer concentrations the dynamics of the single protonated chains can be observed. The NSE measurements were performed at T = 373 K and concentrations $0.04 \leqslant c \leqslant 0.45$ [40, 117].

Figure 64 presents the results of the line-shape analysis for c = 0.18 and c = 0.45. In the first case the polymer relaxation is still determined by the Zimm modes at larger Q-values, while at smaller Q the Rouse modes become dominant. Qualitatively, this behavior is expected for the crossover from unscreened Zimm to enhanced Rouse relaxation. At c = 0.45 the Q-dependence β is

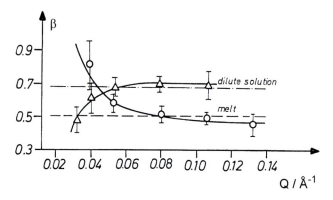

Fig. 64. Single-chain behavior in semi-dilute PDMS/d-chlorbenzene solutions. Line-shape para-meter β as a function of Q at the concentration c = 0.18 and c = 0.45, indicating the occurance of two crossover effects, as predicted by the concept of incompletely screened hydrodynamic interac-tions. (⋯⋯⋯), (– – –) asymptotic Zimm and Rouse behavior, respectively. (Reprinted with per-mission from [117]. Copyright 1984 Elsevier Science Ltd., Kidlington, UK)

inverted. Apparently, Zimm modes are active now over larger distances while the shorter spatial dimensions are governed by Rouse modes. This result, which is incompatible with the assumption that the hydrodynamic interactions are totally screened on larger scales, supports the idea of incomplete screening as supposed by Eq. (133).

Employing the explicitly calculated dynamic structure factors for incomplete screening, the quantitative data analysis demonstrated the consistency of the approach and revealed numerical results for $\xi_H(c)$ and $\eta_H(c)$ [110]. Both quantities were determined from a simultaneous fit of 25 experimental spectra at 5 concentrations, varying only two parameters. Figure 65 presents the concen-tration dependence of $\xi_H(c)$ and compares it with $\xi(c)$, the correlation length for the transition from single chain to collective relaxation in semi-dilute solutions. The magnitudes of both lengths nearly coincide. The concentration dependence $\xi_H(c)$ appears to be closer to the mean field ($\xi_H(c) \sim c^{-1}$) than to the scaling prediction ($\xi_H(c) \sim c^{-3/4}$) and the experimental findings for $\xi(c)$, ($\xi_H(c) \sim c^{0.67}$). However, measurements on more concentrations are needed to determine the exponent. The concentration dependence of $\eta_H(c)$, which controls the crossover from enhanced Rouse to screened Zimm relaxation, is shown in Fig. 65. Since no dependence of $\eta_H(c)$ on the molar mass of the polymer matrix (M ≈ 60.000, 178.000 g/mol) was observed, $\eta_H(c)$ cannot be identified with the macroscopic viscosity of the polymer solution (see Fig. 66).

On the other hand, it was found that the microscopic parameter $\eta_H(c)$ exhibits close similarities to the macroscopic viscosity $\eta(c)/\eta_s$ of a low molecular mass (M ≈ 7.400 g/mol) PDMS/d-chlorobenzene system at 373 K. For that low molar mass the terminal Zimm time τ_Z [see Eq. (80)] is comparable to the time scale of the NSE experiment. Thus, the macroscopic viscosity can relax towards

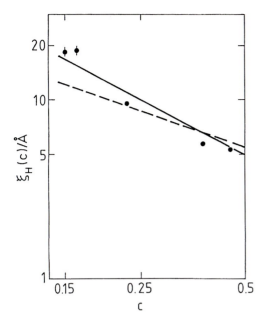

Fig. 65. Concentration dependence of the hydrodynamic screening length $\xi_H (c)$. The *solid line* represents the result of the simultaneous fit, the *dashed line* in correlation length $\xi(c)$ related to the transition from single to many chain behavior. (Reprinted with permission from [40]. Copyright 1984 American Chemical Society, Washington)

equilibrium within the available time window. This coincidence supports the idea that the crossover from enhanced Rouse to screened Zimm relaxation is determined by an effective time-dependent viscosity.

The screening of hydrodynamic interactions was also studied by another group [164], applying the method of "zero average contrast" [30]. This method has the advantage that the scattering experiment reflects the single chain properties, but the intensity can be increased considerably compared with the case [40, 117] where only a small amount of labelled chains are used. The experiments only deal with the concentration c = 0.2 and confirm the earlier observations [40, 117] on the transition from unscreened Zimm to enhanced Rouse relaxation, but do not treat the problem of incomplete screening.

To summarize, the following conclusions arise from the NSE investigation on semi-dilute solutions of homopolymers:

1. Under good solvent conditions the crossover from single-chain relaxation to collective diffusion (many-chain behavior) can be observed by variation of the Q-value
2. The crossover is accompanied by a change in the line shape and in the Q-dependence of the characteristic frequency $\Omega(Q) \sim Q^3 \rightarrow \Omega(Q) \sim Q^2$
3. The crossover appears to be sharp if $\Omega(Q)/Q^2$ is plotted vs. Q
4. Under Θ-conditions the single-chain behavior itself shows, at decreasing Q-values, a crossover from ordinary Zimm relaxation ($\Omega(Q) \sim Q^3$) to an intramolecular microgel mode ($\Omega(Q) \sim Q$), which is due to the occurrence of self-entanglements

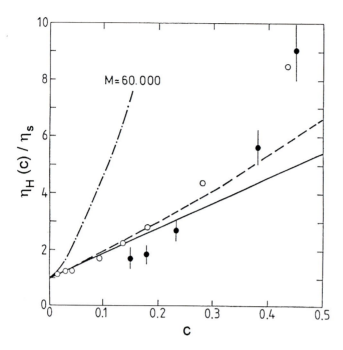

Fig. 66. Concentration dependence of $\eta_H(c)/\eta_S$: ● result of seperate fits; ○ results of viscosity measurements on PDMS solutions ($M_w = 7400$). The result of the simultaneous fit considering the linear term in $\eta_H(c) = \eta_0(1 + [\eta]c + k_H[\eta]^2c^2)$ is given by the *solid line*; the inclusion of a quadratic term leads to the *dashed line*. The *point-dashed line* indicates the macroscopic viscosity for $M = 60000$ g/mol. (Reprinted with permission from [40]. Copyright 1984 American Chemical Society, Washington)

5. The crossover from the single to the many-chain relaxation ($\Omega(Q) \sim Q^2$), which is expected to follow at further decreasing Q-values, has not yet been detected by the NSE technique

6. With respect to the screening of hydrodynamic interactions, one is confronted with the occurrence of a multiple-transition behavior. Instead of the expected crossover from ordinary (unscreened) Zimm to enhanced Rouse relaxation, one observes, at increasing concentrations, additional transitions from enhanced Rouse to screened Zimm and from screened Zimm to enhanced Rouse relaxation. This sequence of crossover effects are highly indicative of an *incomplete screening* of hydrodynamic interactions.

5.5 Semi-dilute Solutions of Polymer Blends and Block Copolymers

5.5.1 Theoretical Outline – The Interdiffusion Mode

Using the random phase approximation (RPA), the coherent scattering intensity $I_{coh}(Q, t)$ of a polymer blend/solvent or a diblock copolymer/solvent system can be written as [169, 170]

$$I_{coh}(Q, t) = A_C e^{-\Omega_c(Q)t} + A_I e^{-\Omega_I(Q)t}. \tag{138}$$

$\Omega_C(Q)$ and $\Omega_I(Q)$ are the characteristic frequencies attributed to the cooperative and the interdiffusion modes, respectively. In the case of the copolymer solvent system, the interdiffusion mode originates from the relative motion of one block with respect to the other.

For both systems A_C becomes zero if the condition of zero average contrast

$$(\rho_M^H - \rho_S)x + (\rho_M^D - \rho_S)(1 - x) = 0$$

or

$$\rho_M^H x + \rho_M^D(1 - x) = \rho_S \tag{139}$$

is fulfilled. In these relations, ρ_M^H, ρ_M^D are ρ_S the coherent scattering lengths per unit volume of the protonated and the deuterated monomers and the solvent, respectively. x is the mole fraction of one of both blocks or homopolymers.

For the special case of a symmetric (x = 1/2) diblock copolymer, A_I is given by

$$A_I^{CP} = n\left(\frac{N}{2}\right)^2 \left(\frac{\rho_M^D - \rho_M^H}{2}\right)^2 (P_{1/2}(Q) - P_T(Q)). \tag{140}$$

n is the number of copolymer chains per volume; N is the total degree of polymerization; $P_{1/2}(Q)$ and $P(Q)$ are the form factors for each block and the total copolymer chain, respectively, which are normalized to unity at $Q = 0$. The derivation of Eq. (140) is based on the additional assumption that both blocks are geometrically symmetrical (same radius of gyration) and that the χ-parameter between the two monomeric species is zero. For the polymer blend/solvent system comparable to the copolymer/solvent system, (140) has to be replaced by

$$A_I^{HP} = nN^2 \left(\frac{\rho_M^D - \rho_M^H}{2}\right)^2 P(Q). \tag{141}$$

In the absence of hydrodynamic, thermodynamic or topological interactions, the corresponding characteristic frequencies $\Omega_I^{CP}(Q)$ and $\Omega_I^{HP}(Q)$ in the regime $Q\langle R_g^2 \rangle^{1/2} < 1$ are derived from Eq. (94) assuming $\mu(Q) = N/\xi$ (Rouse dynamics)

$$\Omega_I^{CP}(Q) = Q^2 \frac{k_B TN}{\xi} \frac{1}{N^2(P_{1/2}(Q) - P(Q))} \tag{142}$$

$$= \frac{6D_0}{\langle R_g^2 \rangle}\left(1 + \frac{3}{8}Q^2 \langle R_g^2 \rangle\right)$$

and

$$\Omega_I^{HP}(Q) = Q^2 \frac{k_B TN}{\xi} \frac{1}{N^2 P(Q)} \tag{143}$$

$$= Q^2 D_0 \left(1 + \frac{1}{3} Q^2 \langle R_g^2 \rangle \right)$$

with $D_0 = k_B T/(N\xi)$.

5.5.2 NSE Results from Polymer Blends and Block Copolymers

NSE measurements at zero average contrast conditions on a symmetric diblock copolymer of H-PS and D-PS dissolved in an appropriate mixture of proto-nated and deuterated benzene are reported [171, 172]. The measurements were performed at different concentrations $c > c^*$. For comparison, the interdiffusion of a corresponding blend of H-PS and D-PS homopolymers dissolved in deuterated benzene was studied, too [171]. Owing to the relatively low molecu-lar masses, only the regime $Q\langle R_g^2 \rangle^{1/2} \lesssim 1$ was accessible, and the internal modes could not be probed.

Figure 67 shows $\Omega_I(Q)/Q^2$ vs. Q for both systems. As expected from Eqs. (142) and (143) their behavior is completely different. One can see that a pro-nounced divergency occurs at small Q-values in the semi-dilute block copolymer solution. If $\Omega_I(Q)/Q^2$ is analyzed in terms of a generalized mobility $\mu(Q)$ [see Eq. (94)], Fig. 68 results from the different concentrations of the diblock copolymer solution. $\Omega(Q)$ varies both with Q and with c. In particular, the Q-dependence is indicative of the non-local character of the mobility and incompatible with the assumption of a pure Rouse type of dynamics. The

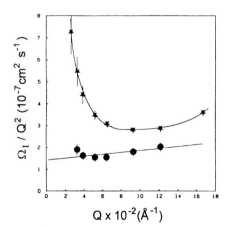

Fig. 67. Experimental variations of the inter-diffusion coefficient $\Omega_I(Q)/Q^2$ as a function of Q at the total polymer concentration $c = 0.34$ g/cm^3 for the two investigated sytems: (*) diblock copolymer PSD-PSH 561/ben-zene + d-benzene; ($\bullet$) mixture of homo-polymers PSH155/PSD425/d-benzene. The *solid lines* are visual aids. (Reprinted with permission from [171]. Copyright 1989 American Chemical Society, Washington)

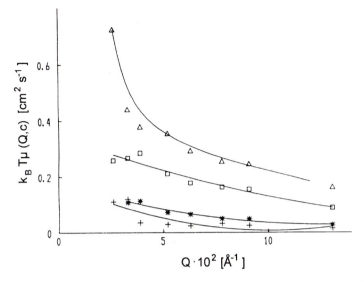

Fig. 68. Mobility $\mu(Q, c)$ as a function of Q [172] at concentrations c in g/cm³. ($\triangle$) c = 15.4 10⁻²; ($\square$) c = 24.6 10⁻²; (*) c = 41.1 10⁻²; (+) c = 48.4 10⁻². (Reprinted with permission from [172]. Copyright 1991 American Chemical Society, Washington)

question, whether hydrodynamics and/or other, e.g. thermodynamic, interactions are responsible for this effect, has not yet been answered. In this connection it would also be helpful to consider the influence of incompletely screened hydrodynamic interactions (see Sect. 5.4.2) on the relaxation of semi-dilute diblock copolymer solutions.

5.6 Collective Diffusion of Tethered Chains

When an A–B diblock copolymer is added to a selective solvent, aggregation analogous to the micellization of surfactants occurs and large areas of tethered chains are achieved. The outer of the polymeric micelles containing the soluble blocks have a structure similar to the outer regions of a multi-arm star. They may also be viewed as polymeric brushes on spherical surfaces. According to de Gennes [173], the dynamics of an adsorbed polymer layer, having a density related to that of a semi-dilute solution of linear chains, is determined by the balance of a restoring force, resulting from the osmotic pressure gradient and a viscous force exerted on the polymer owing to its motion with respect to the background of solvent. As a consequence of the semi-dilute regime the osmotic compressibility E and the viscous force coefficient η/ξ^2 only depend on the correlation length ξ, which scales with the local monomer concentration c as $\xi \sim c^{-3/4}$ [4].

Because of the spherical shape of the PS-PI micelles, the equation of motion for the local displacement $u(r, t)$ of the normal or breathing modes, where all the radical shells or layers move in phase, is given by the radical part of

$$\operatorname{grad}\left[E(r)\operatorname{div}\mathbf{u}\right] = \frac{\eta}{\xi^2(r)}\frac{\partial\mathbf{u}}{\partial t} \tag{144}$$

with $\xi \sim r$ and $E \sim k_B T/\xi^3$ [170]. The solutions of (144) have a simple exponential decay,

$$u_r(t) = u_{r,n}(r)\exp - \{t/\tau_n\}. \tag{145}$$

The eigenvalues $1/\tau_n = a + bn^2$, $n = 0, 1, 2 \ldots$ and $u_{r,n}$, the eigenmode functions for the displacement, are determined by the boundary conditions $u_r(0) = 0$ and $\partial u/\partial r\,(r_{max}) = 0$.

The related dynamic structure factor directly reveals this time dependence and exhibits a multipotential time behavior with a fast initial decay followed by a slowly relaxing tail over longer time periods.

Figure 69 presents typical NSE spectra of such a system [174]. They were obtained from a 2% diblock copolymer (perdeuterated PS and protonated PI blocks) d-n-decane solution. The molecular masses M_w of PS and PI were 10000 and 7500 g/mol, respectively. The solvent selectively dissolves PI but not PS. In

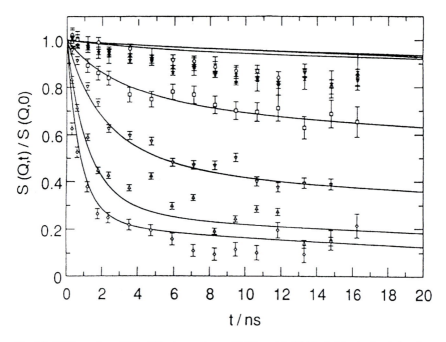

Fig. 69. NSE spectra of 2% diblock copolymer (d-PS and h-PI blocks) in deuterated n-decane. The $Q/\text{Å}^{-1}$ values are: □ 0.026, ▽ 0.032, ∗ 0.038, × 0.051, ◇ 0.064, △ 0.089, ○ 0.115. Experimental data and theoretical dynamic structure for breathing modes are compared (*solid lines*). (Reprinted with permission from [174]. Copyright 1993 The American Physical Society, Maryland)

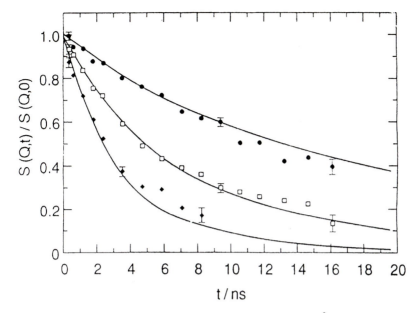

Fig. 70. NSE spectra for 2% linear h-PI in deuterated n-decane at $Q/\text{Å}^{-1}$ values of 0.064 (●), 0.089 (□) and 0.115 (◆). The *solid lines* represent a common fit with the dynamic structure factor of the Zimm model (see Table 1) neglecting possible effects of translational diffusion. (Reprinted with permission from [174]. Copyright 1993 The American Physical Society, Maryland)

addition, when it is perdeuterated, it matches the scattering length density of the PS core so that the contrast is set up. On average, each individual aggregate is composed of 120 chains and has a hydrodynamic radius of 150 Å, whereas the radius of the core is 80 Å. For comparison, the NSE spectra of a corresponding 2% solution of PI homopolymers ($M_w = 17500 \text{ g/mol}$) are also shown (see Fig. 70). While the relaxation of the PI-solution is well described by the Zimm model (see Sects. 5.1.1 and 5.1.2), the dynamics of the PI-corona behaves completely differently. These differences are indicative of an altered physical origin of the underlying motional process. The spectra are characterized by a multiexponential decay function and follow quite well the predictions derived for the breathing mode of chains tethered on a spherical surface.

More detailed investigations with planar brushes, obtained from A-B block copolymers, where one component is able to crystallize on the surface of crystalline PE cores of platelet-like aggregates, have been reported recently [175].

6 Conclusion and Outlook

This article gives a comprehensive review of NSE investigations performed on the large-scale motion of polymers of various architectures in melts, networks and solutions. The method of quasi-elastic small-angle scattering by NSE gives access to the dynamics on the scale of segmental diffusion and allows us to test different microscopic length and time scales and to investigate predictions of universal behavior. The experimental data presented here demonstrate the uniqueness, the power and the efficiency of the NSE technique but also show its limits. These limits, which in particular concern the narrowness of the time window and the time spent on measuring the spectra at different Q-values step by step, will be overcome by the next generation of NSE spectrometers, which are becoming operational now at the Forschungszentrum Jülich and at the ILL in Grenoble.

Acknowledgements We are indebted to the Institute Laue-Langevin (ILL), Grenoble, where the first and for a long time the only operating NSE spectrometer was put into action. This place provided highly stimulating surroundings for all the NSE experiments proposed and performed by us. We are also indebted to F. Mezei, who developed this spectrometer, and to J.F. Hayter and B. Farago, who were involved in all experiments, being responsible for instruments and continuously improving the handling of the spectrometer, pushing it to its limits with respect to the time scale and the energy resolution.

In addition, we acknowledge the important contributions of many colleagues to the various scientific investigations treated by us in the course of years. They are: A. Baumgärtner, K. Binder, J.L. Fetters, J.S. Huang, U. Maschke, R. Oeser and B. Stühn.

Finally, we would like to thank Mrs. I. Brückner-Klein, Mrs. C. Crook, Mrs. I. Nanz, Mrs. I. Serpekian and Mrs. M.L. Schüssler for their technical support and patience preparing this manuscript.

7 References

1. Ferry JD (1980) Viscoelastic properties of polymers. John Wiley & Sons, New York
2. Bird RB, Armstrong RC, Hassager O (1977) Dynamics of polymeric liquids, vol 1. John Wiley & Sons, New York
3. Bird RB, Hassager O, Armstrong RC, Curtis ChF (1977) Dynamics of polymeric liquids, vol 2. John Wiley & Sons, New York
4. de Gennes PG (1979) Scaling concepts in polymer physics. Cornell UniversityPress, Ithaca
5. Doi M, Edwards SF (1986) The theory of polymer dynamics. Clarendon, Oxford
6. Rouse PR (1953) J Chem Phys 21: 1272
7. Zimm BH (1956) J Chem Phys 24: 269
8. de Gennes PG (1971) J Chem Phys 55: 572
9. Edwards SF, Grant JMV (1973) J Phys A 6: 1169
10. Doi M, Edwards SF (1978) J Chem Soc Farad Trans 274: 1789; 274: 1802; 275: 38
11. de Gennes PG (1976) Macromolecules 9: 587 and 594
12. Ewen B, Richter D (1987) Festkörperprobleme 27: 1
13. Richter D, Ewen B (1988) J Appl Cryst 21: 715
14. Richter D (1992) In: Chen SH et al. (eds) Structure and dynamics of strongly interacting colloids and supramolecular aggregates in solutions. Kluwer, Amsterdam, 111
15. Ewen B, Richter D (1995) Polym Symp 90: 131
16. Richter D, Hayter JB, Mezei F, Ewen B (1978) Phys Rev Lett 41: 1484
17. Egelstaff PA (ed) (1965) Thermal neutron scattering. Academic Press, New York
18. Willis BTM (ed) (1973) Chemical applications of thermal neutron scattering, Oxford University Press, London
19. Squires GL (1978) Introduction to the theory of thermal neutron scattering. Cambridge University Press, Cambridge
20. Lovesey SW (1984) Theory of thermal neutron scattering from condensed matter. Clarendon Press, Oxford

21. Sköld K, Price DL (eds) (1987) Methods of experimental physics, vol. 23 A-C. Neutron scattering. Academic. Press, Orlando
22. Allen G, Wright CJ (1975) Int Rev Sci Macromol Sci Phys Chem II, 8: 233
23. Higgins JS, Stein RS (1978) J Appl Cryst 11: 346
24. Maconnachie A, Richards RW (1978) Polymer 19: 739
25. Thomas RK (1979) Mol Spectrosc 6: 232
26. Lechner RE, Riekel C (1981) Z Phys Chem Neue Folge 128: 1
27. Wignall GD (1987) In: Encycl Polym Sci Eng vol 10: Wiley & Sons, New York p 112
28. Sadler D (1988) In: Comprehensive polymer science, vol 1 Allen S, Bevington JC, Booth C, Price C, (eds), Pergamon Press, Oxford p 731
29. Springer T (1972) Quasielastic neutron scattering for the investigation of diffusive motions in solids and liquids. Springer, Berlin Heidelberg New York
30. Bee M (1988) Quasielastic neutron scattering. Adam Hilger, Bristol
31. Higgins JS, Benoit HC (1994) Polymers and neutron scattering. Clarendon Press Oxford
32. Mezei F (1972) Z Phys 255: 146
33. Mezei F (ed) (1980) Neutron spin echo. Springer, Heidelberg Berlin New York
34. Dubois Violette E, de Gennes PG (1967) Physics (USA) 3: 181
35. de Gennes PG (1967) Physics (USA) 3: 37
36. Richter D, Willner L, Zirkel A, Farago B, Fetters LJ, Huang JS (1994) Macromolecules 27: 7437
37. Higgins JS, Gosh RE, Howells WS, Allen G (1977) J Chem Soc Farad Trans II 73: 40
38. Richter D, Baumgärtner A, Binder K, Ewen B, Hayter JB (1981) Phys Rev Lett 47: 109
39. Richter D, Butera R, Fetters LJ, Huang JS, Farago B, Ewen B ((1992) Macromolecules 25: 6156
40. Richter D, Binder K, Ewen B, Stühn B (1984) J Phys Chem 88: 6618
41. Richter D, Ewen B, Farago B, Wagner T (1989) Phys Rev Lett 62: 2140
42. Ewen B, Maschke U, Richter D, Farago B (1994) Acta Polym 45: 143
43. Ewen B, Richter D, Farago B, Maschke U (1993) Prog Colloid Polym Sci 91: 121
44. Richter D, Ewen B (1989) Prog Colloid Polym Sc 80: 53
45. Barlow AJ, Harrison G, Lamb J (1964) Proc R Soc London A282: 228
46. Stühn B, Rennie AR (1989) Macromolecules 22: 2460
47. Pearson DS, Fetters LJ, Graessley WW, Ver Strate G, von Merwall E (1994) Macromolecules 27: 711
48. Pearson DS, Ver Strate G, von Merwall E, Schilling FC (1987) Macromolecules 20: 1133
49. Hess W (1988), Macromolecules 21: 2620
50. Ronca GJ (1983) J Chem Phys 79: 79
51. des Cloizeaux J (1990) Macromolecules 23: 3992
52. Richter D, Willner L, Zirkel A, Farago B, Fetters LJ, Huang JS (1993) Phys Rev Lett 71: 4158
53. de Gennes PG (1981) J Phys (Paris) 42: 735
54. Kremer K, Grest GS (1990) J Chem Phys 92: 5057
55. Kremer K, Grest GS, Carmesin I (1988) Phys Rev Lett 61: 566
56. des Cloizeaux J (1993) J Phys I Fr 3: 1523
57. Higgins JS, Roots JE (1985) J Chem Soc Farad Trans II 81: 757
58. Richter D, Farago B, Fetters LJ, Huang JS, Ewen B, Lartigue C (1990) Phys Rev Lett 64: 1389
59. Butera R, Fetters LJ, Huang JS, Richter D, Pyckhout-Hintzen W, Zirkel A, Farago B, Ewen B (1991) Phys Rev Lett 66: 2088
60. Richter D, Farago B, Butera R, Fetters LJ, Huang JS, Ewen B (1993) Macromolecules 26: 795
61. Graessley WW, Edwards SF (1981) Polymer 22: 1329
62. Kavassalis T, Noolandi J (1988) Macromolecules 21: 2869
63. Colby RH, Rubinstein M (1990), Macromolecules 23: 2753
64. Edwards SF (1967) Proc Phys Soc 92: 9
65. de Gennes PG (1974) J Phys Lett 35: L-133
66. Edwards SF (1988) Proc R Soc Lond A 419: 221
67. Iwata K, Edwards SF (1989) J Chem Phys 90: 4567
68. Lin YH (1987) Macromolecules 20: 3080
69. Colby RH, Rubinstein M, Viovy JL (1992), Macromolecules 25: 996
70. Doi M (1975) J Phys A 8: 959
71. Wu S (1989) J Polym Sci 27: 723
72. Aharoni SM (1983) Macromolecules 16: 1722
73. Fetters LJ, Lohse DJ, Richter D, Witten TA, Zirkel A (1994) Macromolecules 27: 4639

74. Boothroyd AT, Rennie AR, Boothroyd CB (1991) Europhys Lett 15: 715
75. Flory PJ (1986) Principles of polymer chemistry. Cornell University Press, Ithacka, pp 432–494
76. Mark JE, Erman B (1988) Rubberlike elasticity – a molecular primer. John Wiley, New York
77. Queslel JP, Mark JM (1989) In: Allen G, Bevington JC, Booth C, Price C (eds), Comprehensive polymer science. vol 2, Pergamon Press, Oxford, pp 271–309
78. James HM, Guth E (1947) J Chem Phys 15: 668
79. James HM, Guth E (1949) J Polym Sci 4: 153
80. Flory PJ (1944) Chem Rev 35: 51
81. Flory PJ, Rehner J Jr. (1943) J Chem Phys 11: 512; 11: 521
82. Wall FR, Flory PJ (1951) J Chem Phys 19: 1435
83. Flory PJ (1944) Proc R Soc A351: 51
84. Oeser R, Ewen B, Richter D, Farago F (1988) Phys Rev Lett 60: 1041
85. Warner M (1981) J Chem Phys: Solid State Phys 14: 4985
86. Flory PJ, Erman B (1982) Macromolecules 15: 800
87. Erman B, Mark JE (1987) Macromolecules 20: 2892
88. Erman B (1987) Macromolecules 20: 1917
89. Ewen B, Richter D (1992) In: Mark JE, Erman B (eds) Elastomeric polymer networks. Prentice Hall, Englewoods Cliffs
90. des Cloizeaux J, Jannink G (1990) Polymers in solution. Clarendon Press, Oxford
91. Yamakawa H (1971) Modern theory of polymer solutions. Harper and Row, New York
92. Pyum CW, Fixman M (1965) J Chem Phys 42: 3838
93. Akcasu AZ, Gurol H (1976) J Polym Sci Phys Ed: 14, 1
94. Akcasu AZ, Benmouna M, Han CC (1980) Polymer 21: 866
95. Akcasu AZ (1993) In: Brown W (ed) p 1ff Dynamic light scattering – the method and some applications. Clarendon Press, Oxford
96. Zwanzig R (1960) J Chem Phys 33: 1338
97. Mori H (1965) Prog Theor Phys 33: 423
98. Akcasu AZ, Benmouna M, Hammouda B (1984) J Chem Phys 80: 2762
99. Benmouna M, Akcasu AZ (1978) Macromolecules 11: 1187
100. Burchard W, Schmidt M, Stockmeyer WH (1980) Macromolecules 13: 580
101. des Cloizeaux J (1975) J Phys 36: 281
102. Daoud M, Jannink G (1976) J Phys 37: 973
103. Cotton JP, Nierlich M, Boué F, Daoud M, Farnoux B, Jannink G, Duplessix R, Picot C (1976) J Chem Phys 65: 1101
104. Farnoux B, Boué F, Cotton JP, Daoud M, Jannink G, Nierlich M, de Gennes PG (1978) J Phys 39: 77
105. Benmouna M, Akcasu AZ (1978) Macromolecules 11: 1187
106. Nystroem B, Roots J (1982) Progr Polym Sci 8: 333
107. Daoud M, Cotton JP, Farnoux B, Jannink G, Sarma G, Benoit H, Duplessix R, Picot C, de Gennes PG (1975) Macromolecules 8: 804
108. Farnoux B, Daoud M, Decker D, Jannink G, Ober R (1975) J Phys Lett 35: L122
109. Richards RW, Machonnachie A, Allen G (1978) Polymer 19: 266
110. Stühn B (1983) PhD Thesis, Mainz
111. Han CC, Akcasu AZ (1981) Macromolecules 14: 1080
112. Tsunashima Y, Nemeto N, Kurata M (1983) Macromolecules 16: 1184
113. Johnson RM, Schrag JL, Ferry JD (1970) Polymer J 1: 742
114. Warren TC, Schrag JL, Ferry JD (1973) Macromolecules 6: 467
115. Richter D, Ewen B, Hayter JB (1980) Phys Rev Lett 45: 2121
116. Ewen B, Richter D, Hayter JB, Lehnen B (1982) J Polym Sci Polym Lett Ed. 20: 233
117. Ewen B, Stühn B, Binder K, Richter D, Hayter JB (1984) Polym Comm 25: 133
118. Ewen B (1984) Pure Appl Chem. 56: 1407
119. Nicholson LK, Higgins JS, Hayter JB (1981) Macromolecules 14: 836
120. Higgins JS, Ma K, Nicholson LK, Hayter JB, Dodgson K, Semlyen JA (1983) Polymer 24: 793
121. Adam M, Delsanti M (1977) Macromolecules 10: 1229
122. Nemeto N, Makita Y, Tsunashima Y, Kurata M (1984) Macromolecules 17: 425
123. Ewen B (1986) Habilitation Thesis, Mainz
124. Lodge TP, Han CC, Akcasu AZ (1981) Macromolecules 14: 147
125. Akcasu AZ, Benmouna M, Alkhafaji S (1981) Macromolecules 14: 147
126. Osaki K (1973) Adv Polym Sci 12: 1

127. Edwards CJC, Stepto RFT (1986) In: Semlyen JA (ed) Cyclic polymers. Elsevier, London
128. Higgins JS, Dodgson K, Semlyen JA (1979) Polymer 20: 552.
129. Dodgson K, Higgins JS (1986) In Semlyen JA (ed) Cyclic polymers. Elsevier, London
130. Fukatsu M, Kurata M (1969) J Chem Phys 44: 4539
131. Edwards CJC, Stepto RFT, Semlyen JA (1980) Polymer 21: 781
132. Edwards CJC, Bantle S, Burchard W, Stepto RFT, Semlyen JA (1982) Polymer 23: 873
133. Hadjichristidis N, Guyot A, Fetters LJ (1978) Macromolecules 11: 668
134. Hadjichristidis N, Fetters LJ (1980) Macromolecules 13: 191
135. Grest G, Kremer K, Witten TA (1987) Macromolecules 20: 1376
136. Grest G, Kremer K, Milner ST, Witten TA (1989) Macromolecules 22: 1904
137. Daoud M, Cotton JP (1982) J Phys 43: 531
138. Birshtein TM, Zhulina EB (1984) Polymer 25: 1453
139. Birshtein TM, Zhulina EB, Borisov OV (1986) Polymer 27: 1078
140. Benoit H (1953) J Polym Sci 11: 507
141. Burchard W (1983) Adv Polym Sci 48: 1
142. Willner L, Jucknischke O, Richter D, Roovers J, Zhou L-L, Toporowski PM, Fetters LJ, Huang JS, Lin MY, Hadjichristidis N (1994) Macromolecules 27: 3821
143. Grest GS, Fetters LJ, Huang JS, Richter D (1996) Adv Chem Phys 44: 67
144. Zimm BH, Kilb R (1959) J Polym Sci 37: 19
145. Stockmayer WH, Burke JJ (1969) Macromolecules 2: 647
146. Akcasu AZ, Benmouna M, Hammouda B (1982) J Chem Phys 80: 2762
147. Benmouna M, Duval M, Borsali R (1987) J Polym Sci Polym Phys Ed. 25: 1839
148. Benoit H, Hadziioannou G (1988) Macromolecules 21: 1449
149. Richter D, Farago B, Huang JS, Fetters LJ, Ewen B (1989) Macromolecules 22: 468
150. Richter D, Stühn B, Ewen B, Nerger D (1987) Phys Rev Lett 58: 2462
151. Stockmayer WH, Fixman M (1953) Ann NY Acad Sci 57: 334
152. de Gennes PG (1959) Physica (Utrecht) 25: 825
153. Ackerson BJ (1976) J Chem Phys 64: 242
154. Richter D, Farago B, Fetters LJ, Huang JS, Ewen B (1990) Macromolecules 23: 1845
155. Brochard F, de Gennes PG (1977) Macromolecules 10: 1157
156. Brochard F (1983) J Phys 44: 39
157. Tanaka T, Nocker LO, Benedek GB (1973) J Chem Phys 59: 5151
158. Adam M, Delsanti M (1985) Macromolecules 18: 760
159. Chen S-J, Berry GC (1990) Polymer 31: 793
160. Edwards SF, Freed KF (1974) J Chem Phys 61: 1189
161. Freed KF, Edwards SF (1974) J Chem Phys 61: 3626
162. Freed KF, Perico A (1981) Macromolecules 14: 1290
163. Muthukumar M, Edwards SF (1982) Polymer 23: 345
164. Csiba T, Jannink G, Durand D, Papoulov R, Lapp A, Auvray L, Boué E, Cotton JP, Borsali R (1991) J Phys II 1: 381
165. Munch J-P, Candau C, Herz J, Hild GJ (1977) J Phys 38: 971
166. Weill G, des Cloizeaux J (1979) J Phys 40: 99
167. Akcasu AZ, Horn CC (1979) Macromolecules 12: 276
168. Ewen B, Richter D, Farago B, Stühn B (1994) J Non-Cryst Solids 172–172: 1023
169. Benmouna M, Benoit H, Duval M, Akcasu AZ (1987) Macromolecules 20: 1107
170. Benmouna M, Benoit H, Borsali R, Duval M (1987) Macromolecules 20: 2620
171. Borsali R, Benoit H, Legrand J-F, Duval M, Picot C, Benmouna M, Farago B (1989) Macromolecules 22: 4119
172. Duval M, Picot C, Benoit H, Borsali R, Benmouna M, Lartique C (1991) Macromolecules 24: 3185
173. de Gennes PG (1986) CR Acad Sci Paris Ser II 303: 765
174. Farago B, Monkenbusch M, Richter D, Huang JS, Fetters LJ, Gast AP (1993) Phys Rev Lett 71: 1015
175. Monkenbusch M, Schneiders D, Richter D, Farago B, Fetters L, Huang J (1995) Physica B, 213 & 214: 707

Editor: K. Dušek
Received: October 1996

Deformation and Viscoelastic Behavior of Polymer Gels in Electric Fields

Tohru Shiga

Toyota Central Research and Development Laboratories Inc., Nagakute-cho, Aichi-gun, Aichi-ken, 480-11 Japan

"Smart" polymer gels actively change their size, structure, or viscoelastic properties in response to external signals. The stimuli-responsive properties, indicating a kind of intelligence, offer the possibility of new gel-based technology. The article attempts to review the current status of our knowledge of electromechanical effects that take place in smart polymer gels. Deformation and the mechanism of polyelectrolyte gel behavior in electric fields are first studied experimentally and then theoretically. In particular, the swelling or bending is discussed in detail. Particulate composite gels whose modulus of elasticity can vary in electric fields are revealed as a new smart material. The driving force causing varying elastic modulus in electric fields is explained by a qualitative model based upon polarized particles. Finally, applications of the two electromechanical effects are presented.

Advances in Polymer Science, Vol. 134
© Springer-Verlag Berlin Heidelberg 1997

1 Introduction

Polymer gels consist of a crosslinked polymer network and a solvent. Recent advances in polymer gels relate to stimuli-responsive properties [1]. It is well known that polymer gels in a solution swell or deswell when subjected to external stimuli such as changes in pH and temperature, or light. Ionic hydrogels show a discontinuous volumetric change (a first-order phase transition), above a certain threshold of an external stimulus [2, 3]. At the phase transition point, the ratio of the volumes of the collapsed and swollen phases can be as large as several hundred. These phenomena suggest that the size and viscoelasticity of polymer gels may be controlled by the magnitude of the external signals inputted. An electric field is an example of a stimulus that induces a volumetric change in a polymer gel. Because the electric field is the most conventional stimulus from the point of view of signal control, the electric field-responsive behavior may encourage us to fabricate smart gel-based devices such as sensors, actuators, membrane separation devices, drug delivery devices, and vibration isolators.

In this article, I would like to review two types of smart behavior of polymer gels associated with electric fields. One is the deformation of polyelectrolyte gels induced by electric fields (Sects. 2 and 3). The mechanism of deformation will be discussed. The other concerns electrical control of the mechanical properties of gels. Sections 4 and 5 discuss several composite gels whose elastic modulus can vary in electric fields without a volumetric change. Magnetic field-responsive composite gels are also presented. Section 6 describes the applications of the two types of smart behavior of gels.

2 Deformation of Polyelectrolyte Gels in Electric Fields

Electric field-induced deformation of polyelectrolyte gels has attracted much attention because of the property of smartness. If the size and shape of gels can be controlled as we hope, this may open a new door for gel technology. In this Section, studies on electric field-induced deformation of gels will first be surveyed.

2.1 Progress of Research

The electric field-induced deformation of polyelectrolyte gels was first reported by Hamlen et al. in 1965 [4]. They observed that an ionic PVA gel fiber, which was placed touching the anode in a 1% NaCl solution, shrank at the anode side as a result of an applied dc EMF of 5 V. When the polarity of the applied voltage was changed, the deformed gel swelled to the starting size in the absence of an electric field. They explained that the shrinking of the gel at the anode side in electric fields resulted from a change in the chemical structure of the polymer network due to a pH change near the anode. Based upon the work conducted by Hamlen et al., an electrically-activated artificial muscle system using weakly acidic contractile polymer gels sensitive to pH changes was fabricated in 1972 [5]. Tensile forces on the order of 10 g have been measured during the process of shrinking. An artificial muscle using collagen, which is a typical protein polyelectrolyte, has also been proposed [6]. The kinetics of oscillatory tensile forces in collagen membranes under ac electric fields have been analyzed [7]. Similar shrinking deformation of polyelectrolyte gels induced by a pH change in electric fields has been studied by other workers [8, 9].

In 1982, Tanaka et al. attempted an electric field-induced phase transition of a partially hydrolyzed polyacrylamide (PAAm) gel in a mixed solvent of 50% acetone-50% water [10]. The ionic gel showed a phase transition at this acetone concentration. The swollen gel near the phase transition point, which was sandwiched between two electrodes, shrank at the anode under the influence of the dc electric fields. They interpreted the phenomenon as the electric field pushing and squeezing the anode side of the gel. Hirotsu observed a shrinking of ionic gel beads near the phase transition point under dc and ac excitations [11]. Moreover, a unique motion was detected in the shrinking process. The volume of the gel did not decrease monotonically with time. During the shrinking, the gel underwent repeated shrinking and swelling. He pointed out that the complex behavior of the gel was associated with ion drift.

In 1985, a new deformation of a gel in an electric field was reported. Shiga and Kurauchi found that, if the gel is placed at a fixed position separated from two electrodes, a poly(sodium acrylate) (PAANa) gel, which was the same specimen as Tanaka's gel in an NaOH solution, swells at the anode side under a dc electric field [12]. The concentration of NaOH in the surrounding solution strongly affects the aspect of deformation. When the concentration of NaOH is high, the gel swells at the anode side in the presence of an electric field. In contrast, when the concentration is low, the gel shrinks at the anode side. Because the observed swelling or shrinking is a differential deformation, a rectangular gel which is placed parallel to two electrodes bends like a bimetal under dc electric fields. The direction of bending depends on the concentration of NaOH. They explained that the swelling deformation occurs through a change in the osmotic pressure due to the ionic distribution inside and outside the gel under the electric field [13]. The ionic distributions have been calculated using the Nernst-Planck equation [14]. A bending deformation has also been observed in a PAAm gel containing triphenylmethane leuco derivatives when light and an electric field are utilized together [15]. When light is irradiated on the gel, the leuco derivatives are dissociated. Because the gel changes from nonionic to ionic, the gel displays a similar bending behavior when subjected to an electric field. A weakly crosslinked poly(2-acrylamide-2-methylpropane) sulfonic (PAMPS) gel in a solution of n-dodecylpyridinium chloride also shows bending due to a differential shrinking at the anode side, which is caused by the interaction between the negative charges of the polymer network and the positive charges of the surfactant [16].

The electrolysis of water plays an important role in the above-described shrinking and swelling of gels in electric fields. However, because it involves a disadvantage for designing smart devices, i.e., gas formation, the deformation of gels without the electrolysis of water has been studied. A complex microgel of polymethacrylic acid (PMAA) and Ca^{2+} on a polypyrrole (PPy) electrode shrinks in response to a small voltage of less than 1 V [17]. Poly (3-alky-thiophene) (P3AT) is a soluble conjugated polymer. The P3AT gel swollen in a nitromethane solution of $(C_4H_9)_4NClO_4$ also shows an electric field-associated shrinking. The P3AT touching the anode shows not only deformation but an electrochromatic phenomenon due to the action of the electric field [18].

With regard to the response time of the gel, polyelectrolyte gels require seconds to minutes to deform in electric fields. Needless to say, the deformation speed depends on the thickness of the gel and the intensity of the applied field. In 1993, a fast-responsive gel was found by Nanavati and Fernandez. A secretory granule gel particle obtained from beige mice and having a diameter of 3 μm at negative potentials was transparent and swollen within milliseconds of the application of an electric field of 5000 V/cm [19].

2.2 Swelling, Shrinking, and Bending

Understanding the elements which affect the deformation in electric fields is important in designing gel devices. In Sect. 2.2, the aspects of deformation of PAANa gel, which is a typical negatively charged polyelectrolyte gel having ionizable $-COO^-Na^+$ groups, are reviewed. In particular, the deformation of PAANa gel in a solution of monovalent cations is described.

Table 1 shows the observed deformation of a PAANa gel block in dc electric fields. The surrounding solutions used in Table 1 are aqueous hydrochloric acid, water, NaCl solution, or NaOH solution. The swelling ratio of PAANa gel is affected by the pH of the surrounding solution. The swelling ratio in an alkaline region is much larger than that in an acid region. This is because PAANa gel having ionic COONa groups, a strong electrolyte, changes to PAA gel with COOH groups, a weak electrolyte.

I will first discuss the effect of pH on the deformation of PAANa gel which is placed so as to touch the anode. As shown in Table 1, the deformation observed involves only shrinking at the anode side in each solution. It is independent of the pH of the surrounding solution. The ratio of deformation with and without applied fields depends on the nature of the surrounding solution. PAA gel in aqueous hydrochloric acid shrinks slightly under an electric field.

When the gel is placed so as to be separated from the electrodes, the pH of the solution affects the degree of deformation. In alkali and acid regions, both PAANa and PAA gels swell at the anode side on application of electric fields.

Table 1 Deformation of Poly(sodium acrylate) Gel in Electric Fields

Acid Region	Neutral Region		Alkaline Region
in HCl	in water	in NaCl aq.	in NaOH aq.
anode gel cathode			
anode gel cathode			

The deformation in neutral solutions is interesting. A PAANa gel in water shrinks at the anode side, but swells and then shrinks in an NaCl solution. It is suggested that the salt in the solution plays a key role in the swelling.

Let us consider a rectangular gel which is placed so as to be separated from the electrodes. When the gel is placed parallel to the electrodes, the electric field leads to a new deformation, bending, as shown in Fig. 1. The gel bends toward the cathode in HCl, NaCl or NaOH solutions. In contrast, a PAANa gel in water bends toward the anode in an electric field. The bending deformation may be due to the bimetallic principle of anisotropic swelling or shrinking at the anode side. The PAANa gel (70 mm length, 7 mm width, 7 mm thickness) in Fig. 1 bends semi-circularly in 80 s in a 0.02 mol/L NaOH solution on the application of an electric field of 10 V/cm. In the semicircular deformation, the gain in weight is 20%. The increase in weight is proportional to the square of the strain in bending. When a negative field is applied, the gel straightens at the same speed as the bending action. When it returns to the rectangular state, the gel expels water and reverses to its initial size. The deformation speed is proportional to the intensity of the applied field. It also depends on the thickness of the gel. A 1-mm thick gel bends semicircularly in two seconds under 10 V/cm.

In conclusion, the type of deformation induced by an electric field depends on the pH of the surrounding solution, the salt concentration, the position of the gel relative to the electrodes, and the shape of gel.

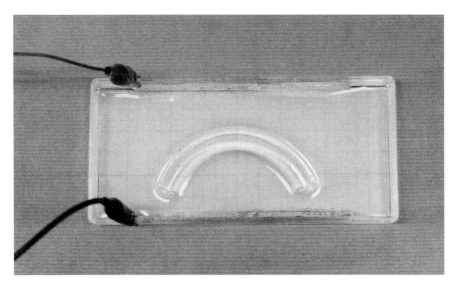

Fig. 1. Bending motion of PAANa gel in dc electric fields. PAANa gel in an NaOH solution swells at the anode side and bends toward the cathode like a bimetal

2.3 Mechanism of Deformation

2.3.1 Swelling

Shiga and Kurauchi have explained that the swelling deformation of a PAANa gel is qualitatively induced by a change in the osmotic pressure based upon a difference in mobile ion concentrations between the inside and the outside of a gel [13]. A change in the osmotic pressure under an electric field has been calculated as follows.

According to Flory's theory [20], the osmotic pressure due to ionic distribution π is given by the following equation,

$$\pi = R T \left(\sum_i C_{in}^i - \sum_j C_{out}^j \right) \tag{1}$$

where R is the Boltzmann constant, T the temperature, C_{in}^i the concentration of mobile ion i in the gel, and C_{out}^j the concentration of mobile ion j in the surrounding solution.

Shiga and Kurauchi have predicted that the swelling deformation of a gel in an electric field is caused by an increase in Flory's osmotic pressure. In order to prove this mechanism, the change in the osmotic pressure under an electric field has been calculated using a simple model for ion transport. The gel and the surrounding solution have been divided into four phases, the A phase (solution at the anode), B phase (gel at the anode), C phase (gel at the cathode), and D phase (solution at the cathode). A mobile cation in a dc field moves toward the cathode, and so the ion transport of a cation from A phase to B phase is expressed using Eq. 2. Eq. 2 has been derived from the fact that the concentration of Na^+ in the A phase decreases linearly with time during deformation.

$$C(t) = C_o(1 - ht) \tag{2}$$

Here C_o is the concentration of a cation in the A phase before the application of an electric field, and h is the transport rate of the cation from the A to B phases.

The concentrations of the cation for the four phases have been calculated as follows.

A phase: $\quad C_A(t) = C_o(1 - ht) \tag{3}$

B phase: $\quad C_B(t) = C_B(1 - ht) + C_o(V_A/V_B)ht(1 - ht) \tag{4}$

C phase: $\quad C_C(t) = C_B(1 - ht) + C_B(V_B/V_C)ht(1 - ht)$

$$+ C_o(V_A/V_C)h^2t^2(1 - ht) \tag{5}$$

D phase: $\quad C_D(t) = C_o + C_B(V_C/V_D)ht + C_B(V_B/V_D)h^2t^2$

$$+ C_o(V_A/V_D)h^3t^3 \tag{6}$$

where C_B is the concentration of the counterion in the gel and V_i the volume of the i phase. In Eq. 2, the second term on the right represents cations which come in from the A phase. In Eq. 3, the second and third terms on the right represent cations from the A and B phases, respectively. The second or third terms on the right side of Eq. 4 represent cations which come in from the B and C phases. The fourth term represents cations from the A phase. In Eqs. 3–6, the assumptions are as follows.

1. The transport rate h is constant everywhere.
2. The concentration gradient in each phase disappears quickly and the concentration becomes uniform.
3. Strong interaction or chemical reaction between mobile cations and negatively charged fixed polyions in gel cannot occur.
4. Cations are not dissipated at the cathode.
5. The effects of diffusion of cations due to a concentration gradient or of the convection of cations and solvent are very small.

The concentrations of a mobile anion in each phase are calculated in the same way as described in the case of the cation.

A phase: $\quad C_A(t) = C_o(1 - ht) + C_o(V_D/V_A)h^3t^3(1 - ht) \tag{7}$

B phase: $\quad C_B(t) = C_o(V_D/V_B)h^2t^2(1 - ht) \tag{8}$

C phase: $\quad C_C(t) = C_o(V_D/V_C)ht(1 - ht) \tag{9}$

D phase: $\quad C_D(t) = C_o(1 - ht) \tag{10}$

Eqs. 7–10 include the same assumptions as those listed above, excluding assumption 4, and the following assumptions are added:

6. Anions in the gel may not exist in the absence of an electric field.
7. Anions vanish from the solution at the anode.
8. The dissipation rate of the anion at the anode is equal to the transport rate of the anion (h).

Because mobile anions cannot exist in the gel before application of a dc electric field, we may discuss only the movement of mobile anions in a solution. The first term on the right side of Eqs. 7–9 displays anions which come in from the D phase. In Eq. 7, the first and second terms on the right contain the dissipation of anions at the anode.

Flory osmotic pressures at the anode and cathode sides have been calculated using Eqs. 3–10. In a system of i cations and j anions, the osmotic pressures are given by a summation of Eqs. 3–10. When $d\pi/dt > 0$ in an electric field, the gel swells. It shrinks when $d\pi/dt < 0$. When π at the anode side becomes larger than that at the cathode side, the gel bends toward the cathode.

Consider the case of the swelling deformation of a PAANa gel (7 mm thick × 70 mm length) in an NaOH solution under a dc electric field of 10 V/cm. In this system, the mobile ions are Na^+, H^+, and OH^-. As mentioned above, the 7 mm-thick gel bends semicircularly in 80 s at 10 V/cm. Assuming that it is given by multiplying the mobility of Na^+ in water by the intensity of the applied field, the drift velocity of Na^+ under 10 V/cm is estimated as follows.

$$V = 5 \times 10^{-4}\,(cm^2\,sec^{-1}\,V^{-1}) \times 10\,(V/cm) = 5 \times 10^{-3}\,cm/sec$$

Thus 140 s are required for Na^+ in the A phase to come across the 7 mm-thick gel. Therefore, Na^+ in the A phase may perhaps move to the C phase during the swelling deformation. This means that the fourth term on the right hand side of Eq. 6 can be neglected.

In Fig. 2, the distributions of mobile ions without and with electric fields are displayed. In each phase, electrical neutrality must be maintained, and so the concentrations in the gel and in the surrounding solution are given by the following equations.

Solution $C_{Na} + C_H = C_{OH}$ (11)

Gel $C_{Na} + C_H = C_P + C_{OH}$ (12)

where C_P is the concentration of the polyion in the gel. In addition, the solubility product of H^+ and OH^- must be kept constant (Eq. 13).

$$C_H \times C_{OH} = 10^{-14}$$ (13)

C_{Na} is much larger than C_H, and the mobile cation is therefore only Na^+ both in the gel and in the solution before application of an electric field. The gel has no mobile anion before application of the field (Fig. 2, left).

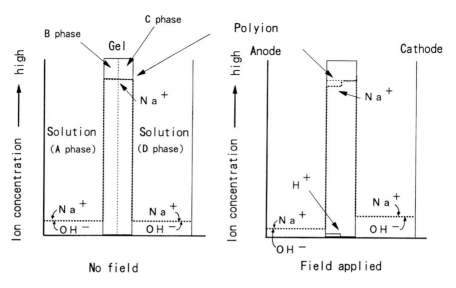

Fig. 2. Ionic distributions in the PAANa gel and in the surrounding NaOH solution with or without dc electric fields. The gel and solution are divided into four phases which are called, in turn, A, B, C, or D phases from the anode side

When an electric field is applied, Na^+ drifts toward the cathode. Therefore the concentration of Na^+ in the A phase decreases with time. The concentration of Na^+ in the B phase also decreases with time (Fig. 2, right). In order to maintain electrical neutrality in the B phase, H^+ is provided from COOH groups. On the other hand, the concentration of Na^+ in the C phase remains constant under an electric field. The concentration of Na^+ in the D phase increases with time. OH^- should be provided to maintain electrical neutrality in each phase. As shown in Fig. 2, we should focus mainly on the concentrations of Na^+ and H^+. The osmotic pressures at the anode and cathode sides at a time t are given by Eqs. 14 and 15.

$$\pi_{anode}(t)/RT = C_{Na,B}(1 - ht) + \alpha\kappa + C_{Na,A} \cdot ht(1 - ht) - 2C_{Na,A}(1 - ht) \quad (14)$$

where $\kappa = C_{Na,B} \cdot ht - C_{Na,A}(V_A/V_B)ht(1 - ht)$

$$\pi_{cathode}(t)/RT = C_{Na,B}(1 - ht) + C_{Na,B} \cdot ht(1 - ht) + 2 C_{Na,A}(V_A/V_B)h^2t^2$$
$$- 2 C_{Na,C} - 2C_{Na,B}(V_B/V_C)ht(1 + ht) \quad (15)$$

where $C_{Na,i}$ represents the concentration of Na^+ in the i phase ($i = A, B$), and α is the dissociation constant of COOH. The second term on the right hand side of Eq. 14 represents the concentration of COO^-H^+ in the B phase, which has been produced from the drift of Na^+. In Eq. 14, a factor of 2 in the third term indicates that the mobile ions are Na$^+$ and OH$^-$. That of 2 in Eq. 15 has the same meaning. In Eqs. 14 and 15, we have assumed that $V_B = V_C$ and $V_A = V_D$.

The osmotic pressures in Eqs. 14 and 15 have been calculated using $C_{Na,B} = 4.41 \times 10^{-5}$ mol/cm^3, $C_{Na,A} = 1.0 \times 10^{-5}$ mol/cm^3, $h = 5.08 \times 10^{-4}$ (s^{-1}), $V_A/V_B = 4$, and $\alpha = 0.1$. $C_{Na,B}$ was determined by the fraction of AANa in the preparation of gel and the swelling ratio of the observed gel. Parameter h was calculated by measuring the time dependence of the concentration of Na$^+$ in the D phase. Fig. 3 indicates the osmotic pressures at the anode and cathode sides calculated from Eqs. 14 and 15. Because the osmotic pressure at the anode side increases with time in an electric field, the gel is found to swell at the anode side. Because the osmotic pressure at the anode side becomes larger than that at the cathode side, the PAANa gel in an NaOH solution bends toward the cathode under the dc electric field. The value of π/RT at $t = 0$ was determined using the experimental data for $C_{Na,B}$ and $C_{Na,A}$, and therefore the calculated results in Fig. 3 may be quite accurate. As shown in Fig. 3, π at the cathode side decreases with time, and so the gel should shrink. Experimentally, there has been no report of shrinkage at the cathode side. It will be interesting to check this point experimentally. The swelling deformation of a PAANa gel in an NaCl solution or in aqueous hydrochloric acid under an electric field has been interpreted using an increase in Flory's osmotic pressure.

Doi and his coworkers have proposed a semiquantitative theory for the swelling behavior of PAANa gels in electric fields [14]. They have considered the effect of the diffusion of mobile ions due to concentration gradients in the gel. First of all, the changes in ion concentration profiles under an electric field have been calculated using the partial differential Equation 16 (Nernst-Planck equation [21]).

$$\frac{\partial c_i}{\partial t} = D_i \frac{\partial^2 c_i}{\partial x^2} - u_i \frac{\partial}{\partial x}(E c_i) \tag{16}$$

where D is the diffusion constant of the mobile ion, u the mobility of the ion, and E the intensity of the applied field. In an equilibrium state, the discontinuity in the ionic concentrations and the electric potential at the gel-solution boundary

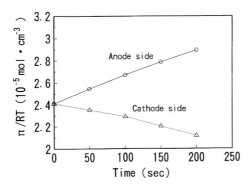

Fig. 3. Osmotic pressures at the anode and cathode sides under dc electric fields. The osmotic pressures are caused by a difference in ion concentrations between the inside and outside of the gel

are determined by the Donnan theory. If an electric field is applied, the system is not in equilibrium, and in general the Donnan theory is not valid. The perturbation from the equilibrium state, however, is very small, and so it has been assumed that the Donnan condition is always satisfied (Eq. 17). They have also considered the continuity in the ionic currents at the gel-solution boundaries (Eq. 18).

$$C_{ig}/C_{is} = \exp\{-z_i(\phi_s - \phi_g)\} \tag{17}$$

$$D_i\left(\frac{\partial c_{is}}{\partial x} + z_i c_{is} \frac{\partial \psi_s}{\partial x}\right) = D_i\left(\frac{\partial c_{ig}}{\partial x} + z_i c_{ig} \frac{\partial \psi_g}{\partial x}\right) \quad \text{for } i = \text{A, B} \tag{18}$$

where C_{ig} and C_{is} are the concentrations of mobile ion i in the gel and in the solution, z the charge of mobile ion i, and ϕ the nondimensional electrical potential. It is assumed that the diffusion constant of ion i in the solution is equal to that in the gel. Finally, they have combined the calculated results of the ion concentrations with Flory's theory (Eq. 1) and have obtained a time dependence of π. The profiles of ion distributions at a point x in gel are described at a time t. The obtained results have predicted that the swelling behavior is governed by the change in osmotic pressure due to a difference in ion concentrations between the inside and the outside of the gel.

   In addition, these authors have discussed the deformation of a PAANa gel placed in contact with the cathode in a dc electric field. Because there is no interaction between the gel and the cathode, the gel swells at the anode side, as observed in the gel separated from the electrodes. It is concluded that the gel also swells at the anode side under an electric field.

2.3.2 Shrinking

The shrinking deformation of a PAANa gel touching the anode in water is first discussed. This may be the same as that observed in a mixture of 50% acetone-50% water by Tanaka et al. [10]. These workers have proposed that the shrinking is induced by electrostatic forces between the anode and the negative charges of polyions in the gel. This explanation, however, is questionable, because Tanaka with Hirose have calculated a change in pH gradient in the gel under a stationary electric field using Eq. 16 and have proposed that the shrinking is induced by a pH gradient in the gel later on [22]. As the pH at the anode side decreases with time, a change in the chemical structure of PAANa to PAA occurs in an electric field, and a drastic volume shrinkage will then occur.

   The shrinking of a PAANa gel touching the anode in an NaCl solution is the same phenomenon as that reported by Hamlen et al. When an electric field is applied, Cl^- ions concentrate near the anode. Therefore, the pH at the anode decreases with time, and the pH change undoubtedly leads to a change from a PAANa gel to a PAA gel.

The shrinking of a PAANa gel touching the anode in an NaOH solution has been analyzed by Doi et al. [14]. They have calculated the osmotic pressure at the anode side using Eq. 16. At the anode, H^+ ions are produced by the electrolysis of water, and this suppresses the dissociation of carboxyl groups near the anode. As a result, π at the anode side decreases very quickly. Thus the gel shrinks when it is kept in contact with the anode.

There are two explanations for the small shrinkage in hydrochloric acid. One is that the deformation may be caused by the electrostatic forces between the anode and the negatively charged polyions in the gel. The other relates to the screening effect of $COO^- H^+$ by Cl^- which comes in from the D phase. As a large number of Cl^- ions restrain interactions between $COO^- H^+$, the gel will shrink.

The deformation of a PAANa gel in water, placed to separate two electrodes, has been studied by Shiga et al. [23]. UV spectra of a PAANa gel at the anode side have been measured with and without an electric field. The absorption of COONa decreases with time in an electric field, while that of COOH increases. The shrinking has been concluded to occur through a change in the chemical structure of PAANa to PAA.

In Sects. 2.2 and 2.3, the electric field-induced deformation of a PAANa gel and its mechanism in a solution of a monocation have been discussed. The observed swelling or shrinking is associated with the drift of mobile ions. The swelling may be caused mainly by increasing osmotic pressure due to a difference in mobile ion concentrations between the inside and the outside of the gel. In contrast, a change in the pH gradient plays a key role in the shrinking. It may lead to a change in the chemical structure of COONa to COOH, with a drastic change in volume.

3 Complex Deformation of Ionic Gels in Electric Fields

In this Section, we describe two complex deformations due to electric field-associated swellings. One is the bending of a hybrid gel consisting of a PVA-PAA gel rod and a PVA film in dc fields. The hybrid gel has been used to fabricate a smart gel finger. The other is the vibration of PVA-PAA gel film in ac fields. This new deformation suggests a gel having a fast response on the order of 100 ms.

3.1 Bending of a Hybrid Gel

On freezing and subsequent thawing, an aqueous solution of PVA changes to a PVA gel [24]. According to X-ray diffraction and pulsed NMR studies [25, 26], the gelation of PVA solution is based upon microcrystallization of PVA chains as crosslinked domains. When the freezing and thawing process is repeated, the elastic modulus of the prepared gel increases with the number of repetitions. The obtained PVA gel has a relatively high tensile strength and high elasticity. However, the PVA gel exhibits no response to an electric field, because it has no ionizable groups. "Ionic PVA gel" (PVA-PAA gel) has been synthesized by repeatedly freezing and thawing a mixture of PVA and PAA [27]. The PVA-PAA gel in an NaOH solution under a dc electric field swells at the anode side and bends toward the cathode like a PAANa gel [28]. However, it has recently been found that a PVA-PAA gel bends toward the anode when it is covered with a thin PVA gel film having a thickness of 500 μm [29]. In this Section, we will discuss why such a hybrid gel bends toward the anode in an electric field.

Figure 4 indicates the relationship between strain in bending and weight change of a gel in an electric field. Although the gain in weight for the hybrid gel is smaller than that for a PVA-PAA gel, the bendings of both PVA-PAA and hybrid gels are found to be caused by a differential swelling. The main difference between them is the direction of bending. Shiga et al. have attempted to explain this complex behavior by calculating the osmotic pressure associated with the concentrations of mobile ions [29]. They have considered that the swelling behavior of the hybrid gel in an electric field may perhaps occur at the cathode side.

Let us start with the equations for the osmotic pressures at the anode and cathode sides (Eqs. 14 and 15). These are as follows.

$$\pi_{anode}/RT = C_B(1 - ht) + \alpha\kappa + C_{Na,A}(V_A/V_B)h't(1 - ht)$$
$$- 2C_{Na,A}(1 - h't) \tag{19}$$

$$\pi_{cathode}/RT = C_B(1 - h't) + C_B ht(1 - h't) + 2C_{Na,A}(V_B/V_C)hh't^2$$
$$- 2C_{Na,A} - 2C_B(V_B/V_C)h't(1 + ht) \tag{20}$$

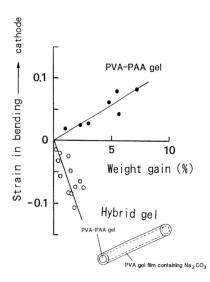

Fig. 4. Strain in bending versus weight change in hybrid gel (○) and PVA-PAA gel (●). The weight of both gels increases in an electric field

where C_B is the concentration of the counterion in the gel. In Eqs. 19 and 20, h' represents the transport rate of Na^+ from solution to gel or from gel to solution. The fraction of water in the PVA gel film of the hybrid gel is 78%, while that in a PVA-PAA gel rod is 93%. The authors have therefore predicted that h' is smaller than h (transport rate of Na^+ from gel to gel).

Figure 5 shows the calculated results for the cases of $h = h'$ ($\bigcirc$, $\bullet$) and of $h' = h/5$ ($\square$, $\blacksquare$). The osmotic pressures have been calculated using $C_{Na, A} = 2.0 \times 10^{-5}$ mol/cm^3, $C_B = 8.0 \times 10^{-5}$ mol/cm^3, and $h = 5.08 \times 10^{-4}$ s^{-1}. C_B was obtained from the fraction of PAA in the preparation of the gel and the swelling ratio of the PVA-PAA gel. We used the same value of h in Sect. 2.3.1. When $h' = h$, indicating a PVA-PAA gel, π at the anode side ($\bigcirc$) increases with time in an electric field. The parameter π at the cathode ($\bullet$) increases slightly. Thus, the gel should swell and bend toward the cathode. The result is in agreement with the experimental results for a PVA-PAA gel. When $h' = h/5$, the osmotic pressure at the anode side ($\square$) decreases with time while that at the cathode ($\blacksquare$) increases. Because the gain in the osmotic pressure at the cathode $|\pi(t)/\pi(0)|$ becomes larger than that at the anode, the gel should swell at the cathode side and bend toward the anode. The case of $h' = h/5$ may qualitatively correspond to the deformation of the hybrid gel. The relationship $h' = h/5$ means that the PVA gel film prevents the transport of mobile ions from solution to gel. Therefore, to summarize, the bending behavior of the hybrid gel is based upon a swelling at the cathode side due to a difference in the concentration between mobile ions on the inside and those on the outside of the gel.

3.2 Vibration of Gel Film

If the bending of a gel in an electric field is regarded as a three-point mechanical bending of a uniform cantilever, the strain in bending is given by $\varepsilon = 6DY/L^2$ (D: thickness of gel, L: length of gel before application of the electric field, Y: deflection of bending). The strain in bending is experimentally proportional to

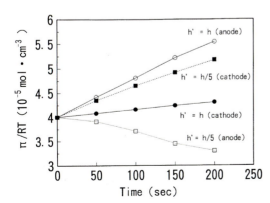

Fig. 5. Time dependence of osmotic pressures at the anode and cathode sides in the hybrid gel. The parameters h and h' represent the transport rates of Na^+ from solution to gel and from gel to gel, respectively. When $h' = h/5$, the gel shows anisotropic swelling at the cathode side and bends toward the anode side

the difference in the osmotic pressures between the anode and the cathode sides $\Delta\pi$. Therefore, the relationship between $\Delta\pi$ and ε is described as follows.

$$\Delta\pi = \pi_{\text{anode}} - \pi_{\text{cathode}} = \varepsilon E = 6DYE/L^2 \tag{21}$$

where E is the elastic modulus of gel. When the electrical power applied is constant, in other words, when $\Delta\pi$ is constant, a thinner gel shows a larger deflection or a higher response. It will therefore be expected that a thin gel responds to ac excitation. It is effective for a gel to become slender. The electric field-associated deformation of PVA-PAA gel with a diameter of the order of 1 mm has been studied [30]. The PVA-PAA gel of 1 mm diameter bends semicircularly and straightens in an electric field of 0.5 Hz. In this section, we will discuss the deformation of PVA-PAA gel films with thicknesses of less than 1 mm under ac excitation.

Figure 6a shows an experimental setup for the measurements of deformation of the PVA-PAA gel films (10 mm width × 40 mm length) under ac electric fields. The gel film fixed on a load cell was immersed in a 10 mM Na_2CO_3 solution. An ac electric field of 15 V/cm and a frequency of 0.1–20 Hz was then applied across the gel film between two electrodes. The deformation behavior of the gel films in electric fields was monitored by a video camera [31]. Figure 6b indicates the deformation of a 0.41 mm-thick gel under ac excitation. Like the PAANa gel, the PVA-PAA gel film shows a bending-straightening motion when subjected to electric fields of less than 1 Hz. However, it begins to wave or vibrate under signals of 2 to 5 Hz. With a 2 Hz signal, two nodes were observed at $x = 0$ and at $x = 0.75\,L$ (x: distance from the fixed point, L: length of the film). There are two antinodes with a deflection of 2 mm at $x = 0.52\,L$ and $x = 1.0\,L$. As seen from Fig. 6b, the number of antinodes increases with the frequency of the applied field.

The experimental results suggest that the vibration of the PVA-PAA gel films with ac signals is derived from the bending. Here, attempts are made to analyze the vibration of gel films to support the view that this is caused by a differential swelling. As already mentioned, the bending of the gel is regarded as a mechanical bending vibration of a uniform cantilever beam due to sinusoidally varying forces. Let us begin from a differential equation for the deflection of the bending vibration of a uniform cantilever beam with a clamped end and a free end [32, 33]. If the effects of shear deformation and rotatory inertia are neglected, the differential equation is

$$EI(\mathrm{d}^4w/\mathrm{d}x^4) + \omega^2 mw = F(x, t) \tag{22}$$

where W is the deflection, E the Young's modulus of the gel, I the moment of inertia of the beam cross-section, m the beam mass per unit length, and $F(x, t)$ the external force [$F(x, t) = C \sin 2\pi\, ft$].

The solution to Eq. 22 is generally expressed by $w(x, t) = \Sigma w(x)q(t)$, so that $w(x)$ can be calculated from Eq. 22 at $F(x, t) = 0$.

$$EI(\mathrm{d}^4w/\mathrm{d}x^4) + \omega^2 mw = 0 \tag{23}$$

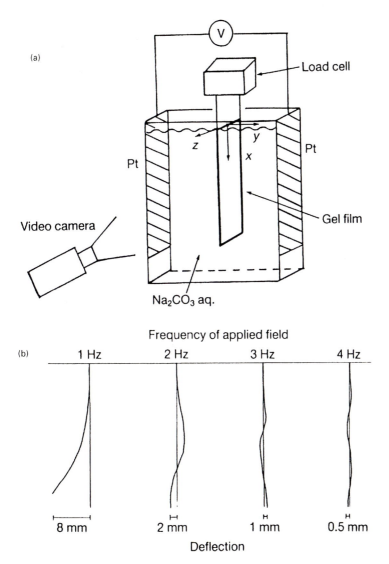

Fig. 6a. Schematic illustration of deformation measurement of PVA-PAA gel film in electric fields.
b Deflection curves of PVA-PAA gel film under sinusoidally varied electric fields

Eq. 23 is the so-called Bernoulli-Euler beam equation. The solution to this fourth-order equation contains four constants and is written in the form,

$$w(x) = C_1 \cosh \beta x + C_2 \sinh \beta x + C_3 \cos \beta x + C_4 \sin \beta x$$

$$\beta = \omega^4 m / EI \tag{24}$$

For a cantilever beam with x separated from the fixed end, the boundary conditions are $w(0) = 0$, $w'(0)$, $w''(L) = 0$ and $w'''(L) = 0$. Using these boundary condtions, we obtain the frequency equation

$$\cosh \lambda \cos \lambda = -1, \qquad \lambda = \beta L \tag{25}$$

where $\lambda_1 = 1.8751$, $\lambda_2 = 4.6941$, and $\lambda_i = (2i - 1)\pi/2 \, (i \geqslant 3)$. Using the eigen-values λ, the natural frequency is determined from Eq. 25.

$$\omega = \lambda^2 (EI/m)^{1/2}/L^2 \tag{26}$$

In the mechanical bending, when $n = 2$, there are two nodes at $x = 0$ and $x = 0.774$ in the deflection curve $w_2(x)$ of Eq. 26. As shown in Fig. 6b, with a 2 Hz signal, the second node has been located at $X = 0.75\,L$. There is a second antinode at $x = 0.52\,L$. So the deflection curve of the gel film with a 2 Hz signal is similar to $w_2(x)$.

The cross-section of the gel film is a rectangular plate of the dimension of bt. Here b and t represent the width and thickness of the film, respectively. Therefore the moment of inertia of the beam cross-section is given by Eq. 27.

$$I = \int_{-\frac{t}{2}}^{\frac{t}{2}} y^2 b \; \mathrm{d}y = bt^3/12 \tag{27}$$

Because the beam mass per unit length m is calculated from $m = \rho bt$ (ρ = density), inserting Eq. 27 into Eq. 26 results in

$$\omega = \lambda^2 t(E/12\rho)^{1/2}/L^2 \tag{28}$$

Equation 28 states, that ω is proportional to t. The effect of the thickness of the gel film on the frequency of the first resonance mode has been investigated. When the buoyancy is taken into account, the experimental results have quantitatively followed Eq. 28. It has been found that the buoyancy plays an important role in the occurrence of the electric field-associated vibration of gel film. The vibration of the gel film in an electric field has thus roughly analyzed as a mechanical bending vibration of a uniform cantilever beam.

4 Electroviscoelastic Behavior of Composite Gels

Sounds and vibrations in electrical appliances or vehicles are in some cases unpleasant. If we can reduce these sounds and vibrations as we desire, we will have a more happy and comfortable life in the next century. Sections 4 and 5 discuss a new smart polymer gel for actively reducing sounds and vibrations. The smart gel can vary its elastic modulus in an electric field.

4.1 Principle of Electrical Control of Viscoelasticity

Some suspensions of polymer particles with high dielectric constants dispersed in nonconducting oils stiffen rapidly when subjected to electric fields on the order of 1 kV/mm [34–37]. Such rheological behavior, known as the electro-rheological effect (ER effect), was first reported by Winslow in 1947 [38]. The ER effect is explained step by step as follows. The dispersed particles polarized electrically as a result of the action of electric fields. The electrically polarized particles move and align between the electrodes. Many chains of the particles formed between the electrodes produce an increase in viscosity. Although the phenomenalogical mechanism of the ER effect is generally accepted, it has not, to date, been determined how the polarization of the dispersed particles occurs at a molecular level. The first model for polarization has been proposed by Klass and Martinek [39]. They investigated the ER effect of silica particles in ac electric fields and suggested that the effect is based upon a distortion of the electrical double layer. This mechanism has been widely accepted by other workers [40].

In the 1980s, some models for polarization of particles were proposed. Stangroom pointed out that the formation of a water bridge at the interfaces of dispersed particles is important in an ER fluid of poly-electrolyte particles [41]. The mechanism of the water bridge was discussed theoretically [42]. In the case of the ER effect of semiconducting particles, the formation of dipoles in particles is governed by surface charge migration or bulk charge migration [43]. Further-more, a point dipole approximation model has been used in computer simula-tion for the dynamics of the ER effect [44–47]. As mentioned above, the mechanism of polarization at a molecular level is one of the outstanding problems. It can, however, be broken down roughly into polarization at the interface between a particle and oil and into polarization in a particle.

It is generally thought that the ER effect happens only in nonconducting oils. Here an ER effect in solid-like matrices such as polymer gels will be discussed. The nature of the ER effect in polymer gels will be explained using the point dipole model in [44].

Polarization in the point dipole model occurs not at the surface of the particle but within it. If dipoles form in particles, an interaction between dipoles occurs more or less even if they are in a solid-like matrix [48]. The interaction becomes strong as the dipoles come close to each other. When the particles contact each other along the applied electric field, the interaction reaches a maximum. A balance between the particle interaction and the elastic modulus of the solid matrix is important for the ER effect to transpire. If the elastic modulus of the solid-like matrix is larger than the sum of the interactions of the particles, the ER effect may not be observed macroscopically. Therefore, the matrix should be a soft material such as gels or elastomers to produce the ER effect.

Let us consider the elastic modulus induced by the ER effect. Jordan et al. have directly measured shear forces between adjacent particles in a chain of particles under electric fields, which is the interaction of particles, and have obtained shear forces on the order of 1×10^{-5} N [49]. My laboratory team has estimated the macroscopic increase in elastic modulus of the matrix due to the ER effect using a simple model (see Sect. 4.2). When the shear force is 2×10^{-5} N, it is expected that the increase in the elastic modulus induced by an electric field is 4 kPa. The increase of 4 kPa has been calculated using Eq. 34 with the particle volume fraction 0.3, the particle diameter 50 µm, and the intensity of the applied field 2 kV/mm. It has therefore been concluded that behavior similar to the ER effect appears in a solid-like matrix having an elastic modulus on the order of 1 kPa.

Based upon the calculated results, we have prepared composite gels composed of silicone gels having various elastic moduli and poly-methacrylic acid cobalt(II) salt (PMACo) particles. The shear modulus of the composite gels with or without and electric field has been measured [50]. Note that the silicone gels used do not include a solvent. Figure 7 shows the effect of volume fraction of PMACo particle on the shear modulus of the composite gels with a shear modulus of 4 kPa. The PMACo particles used have an absorbed water content of 12.7 wt%. When the volume fraction of particle is more than 25 vol%, the shear storage modulus increases on the application of a dc electric field of 5 kV/mm. That is, an effect similar to the ER effect has been demonstrated (ER gel obtained). According to our microscopic observation, the ER effect has been detected in gels in which the dispersed particles have already formed many migration paths before application of an electric field. The aggregation of dispersed particles on application of an electric field has not been observed.

From the experimental results, the ER effect in polymer gels is explained as follows (Fig. 8). When an electric field is applied, the particles electrically bind together and cannot slip past each other. Larger shear forces are needed in the presence of an electric field. Thus, the electric field apparently enhances the elastic modulus of the composite gel. The difference in ER effects between an oil and a gel is that the polarized particles necessarily cannot move between the electrodes to produce the ER effect in a gel. In order for the ER effect to occur, it is important to form migration paths before application of an electric field.

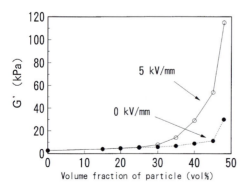

Fig. 7. Effect of volume fraction of dispersed particles in a gel on the shear modulus of the gel under electric fields. The sample used was a silicone gel containing PMACo particles. The particles were dispersed at random in the gel

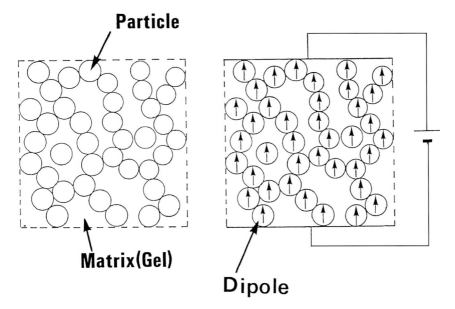

Fig. 8. Schematic illustration of the ER effect in a polymer gel. The paths of the particles have been formed before application of an electric field

Concerning the effect of dispersion of the particles, straight migration paths as shown in Fig. 9 are effective to show a large ER effect.

Because an electric field also involves a magnetic field, it is suggested that the materials which increase elastic modulus in external magnetic fields (MR gel) can be prepared using particles which can polarize in a magnetic field. An MR gel will be discussed in Sect. 5.1.

4.2 Theoretical Approach

4.2.1 Viscoelasticity in Electric Fields

In this section, we discuss theoretically the influence of microscopic interaction between polarized particles on the macroscopic mechanical properties of the composite gel, in particular, the elastic modulus.

Let us consider a hard sphere in a continuous matrix under an electric field. When the relative dielectric constant of particle ε_2 is larger than that of matrix ε_1, a point dipole in the particle is formed by application of an electric field. According to the classical theory [51], the point dipole moment is given by

$$\mu = 4\pi \, r^3 \, \varepsilon_0 \, \varepsilon_1 \, \kappa \, E \tag{29}$$

$$\kappa = (\varepsilon_2 - \varepsilon_1)/(\varepsilon_2 + 2\varepsilon_1)$$

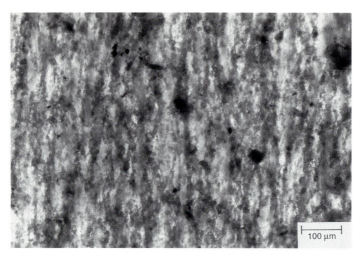

Fig. 9. Photograph of a dispersion of particles in a gel. The dark parts represent the particles aligned uniaxially

where r is the radius of the particle, ε_0 the permittivity in vacuo ($= 8.854 \times 10^{-12}$ F/m), and E the intensity of the applied field. Equation 29 is an expression obtained formally for the volume polarizability of a particle.

Next consider an interactive force between two dipole particles separated by R. The dipole–dipole interaction F is given by the following equation [52].

$$F(R, \theta) = R_0\{(2r/R)^4[(3\cos\theta - 1)e_r + \sin(2\theta)e_\theta]\} \qquad (30)$$

where e_r and e_θ are vectors. When two particles are aligned along the applied electric field and are in contact with each other, Eq. 30 reduces to Eq. 31.

$$F = (3/2)\pi\, r^2\, \varepsilon_0\, \varepsilon_1\, \kappa^2\, E^2 \qquad (31)$$

Let us consider a cubical gel having many straight migration paths of dispersed particles which remain parallel to the direction of the applied field, and discuss the deformation of the cube (length L, volume fraction of particle C) by a small shear strain $\gamma(= x/L)$ in a direction perpendicular to the applied field. We assume that electrical interactions between dispersed particles in a path are based only upon an interaction between two adjacent particles and that interactions between paths of particles are negligible. If macroscopic mechanical properties such as storage and loss moduli are estimated easily by multiplying the electrostatic force between adjacent particles at short range in the path, F being the number of the path of the particle, Eq. 32 is obtained.

$$F\sin\theta \cdot n/L^2 = \gamma(G_E - G_0) \qquad (32)$$

where G_E and G_0 are elastic moduli with and without an electric field, respectively. The parameter n is given by Eq. 33,

$$n = \frac{\text{vol of particle in a cube}}{\text{vol of particle in one path}} = 3CL^2/2\pi r^2 \tag{33}$$

$\sin \theta = x/(x^2 + L^2)^{1/2}$; therefore, an increase in elastic modulus due to an applied electric field ΔG is given as follows:

$$\Delta G = (9/4)C\,\varepsilon_1\,\kappa^2 E^2 \tag{34}$$

Microscopic interaction of the ER effect is correlated with macroscopic mechanical properties through Eq. 34. This states that ΔG is proportional to C, ε_1 or E^2 and depends on ε_2. When ε_2 increases, κ^2 increases and then reaches a maximum. ΔG may perhaps become saturated.

4.2.2 Viscoelasticity in Magnetic Fields

Let us discuss the viscoelasticity of a cubical gel having many straight migration paths of dispersed particles under a magnetic field. Let us start with the induced magnetic dipole moment of a sphere in an unbounded continuous medium which is not influenced by external magnetic fields. The magnetized particle may be considered as a point dipole of magnetic moment M [53].

$$M = (4/3)\pi r^3 \chi H \tag{35}$$

where r is the radius of a sphere, χ the magnetic susceptibility of a sphere per unit volume, and H the intensity of the applied magnetic field.

The magnetostatic potential about an isolated pair of spheres with dipole moments of M_1 and M_2 will now be considered. M_1 and M_2 are vectors. The magnetostatic potential ϕ between M_1 and M_2 separated by distance R is given by Eq. 36.

$$\phi = 1/(4\pi\mu_0 R^3)[M_1 \cdot M_2 - (1/3)(M_1 \cdot R)(M_2 \cdot R)] \tag{36}$$

Here μ_0 is the magnetic permeability. When the two spheres are aligned in the direction of the external magnetic field and touch each other. ϕ is given by Eq. 37.

$$\phi = -M_1 \cdot M_2/2\pi\mu_0 R^3 \tag{37}$$

Because the magnetostatic force between two adjacent spheres at a short range is calculated by the relation, $F = -d\phi/dR$.

$$F = 3M_1 \cdot M_2/2\pi\mu_0 R^4 = -\pi r^2 \chi^2 H^2/6\mu_0 \tag{38}$$

In Eq. 38, the negative sign represents the attractive force of F.

Let us now discuss the shear modulus of the composite gel under a magnetic field which is applied parallel to the path of a dispersed particle. Shear strain is applied so as to cross the path of a particle. The assumptions are as follows:

1. The interaction of particles in a path is governed by the magnetostatic forces of two adjacent particles.
2. The interactions between the paths are very small.

Thus, ΔG reduces to Eq. 39 using Eq. 32.

$$\Delta G = C \chi^2 H^2 / \mu_0 (\gamma^2 + 1)^{1/2} \tag{39}$$

In Eq. 39, ΔG depends on $C \cdot \chi$ is influenced by H. However, when χ is independent of H, ΔG is proportional to the square of H. ΔG remains constant over a small range of γ.

5 Viscoelastic Properties of Composite Gels in Applied Fields

5.1 ER or MR Gels

We have called composite gels which vary in elastic modulus in electric or magnetic fields ER or MR gels. In this section, the viscoelastic properties of some ER or MR gels are presented

The dynamic viscoelasticity of composite gels composed of a silicone gel and PMACo particles (diameter ca. 75 μm) containing a small amount of water has been investigated under the influence of electric fields [54]. Figure 9 indicates the dispersion of the PMACo particles in the gel. The PMACo particles align uniaxially and form many migration paths before application of an electric field. The materials were prepared by heating a prereaction solution of silicone gel containing the suspended PMACo particles under an electric field. Figure 10 indicates the shear storage and loss moduli of the gel in various fields. Electric fields up to 5 kV/mm were applied parallel to the paths of the particles. As expected from Eq. 34, $\Delta G'$ and $\Delta G''$ are proportional to the square of the field intensity. The influence of electric fields is found to be reflected in $\Delta G'$ rather than in $\Delta G''$. The effects of volume fraction and of dielectric constant of the particle on $\Delta G'$ and $\Delta G''$ have also been investigated. When ε_2 increases, $\Delta G'$ and $\Delta G''$ initially gain upon the application of an electric field and then attain a maximum. The results obtained have suggested that the viscoelastic behavior of the PMACo particle/silicone gel in an electric field is qualitatively explained by the simple model of the point dipole. The ER effect of the particulate gel in electric fields has been investigated in the compression mode. Compressive strain has been applied parallel to the paths of the PMACo particles. An increase in the compressive modulus induced by an applied electric field obeys an equation similar to Eq. 34.

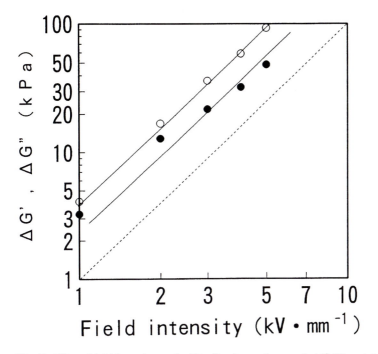

Fig. 10. Effect of field intensity on the ER effect in a polymer gel. $\Delta G'$ ($\bigcirc$) and $\Delta G''$ ($\bullet$) represent increments in shear storage and loss moduli of the gel induced by applied electric fields. The dotted line displays the relationship $\triangle G = kE^2$ (k constant, E field intensity)

The dynamic viscoelasticity of particulate gels of silicone gel and lightly doped poly-p-phenylene (PPP) particles has been studied under ac excitation [55]. The influence of the dielectric constant of the PPP particles has been investigated in detail. It is well known that the dielectric constant varies with the frequency of the applied field, the content of doping, or the measured temperature. In Fig. 11 is displayed the relationship between an increase in shear modulus induced by ac excitation of 0.4 kV/mm and the dielectric constant of PPP particles, which was varied by changing the frequency of the applied field. $\Delta G'$ increases with ε_2 and then reaches a constant value. Although the composite gel of PPP particles has dc conductivity, the viscoelastic behavior of the gel in an electric field is qualitatively explained by the model in Sect. 4.2.1, in which the effect of dc conductivity is neglected.

An interesting phenomenon has been observed in an ER gel of iodine-doped poly(3-hexylthiophene) (P3HT) particles with dc conductivities on the order of 10^{-7} S/cm. The applied electric fields usually enhance the elastic modulus of ER gels. On the other hand, the elastic modulus of the P3HT particle system decreases in an electric field [56]. It has been suggested that the electroviscoelastic behavior of the P3HT particle system may be caused by softening of the P3HT particles themselves in an electric field.

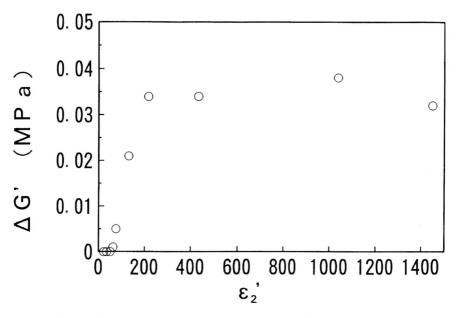

Fig. 11. Effect of dielectric constant of dispersed particle on the ER effect in a polymer gel. The increment in shear storage modulus induced by an ac electric field of 0.4 kV/mm is plotted as a function of the real part of the dielectric constant of the particle

The elastic modulus of laminated composite plate in which an ER silicone gel of carbonaceous particles is sandwiched between two PVC sheets also changed under the influence of an electric field. It was found that an electric field of 1.17 kV/mm caused a gain in the elastic modulus of the gel of 13% [57].

The dynamic viscoelasticity of an MR gel composed of a silicone gel and iron particles has been studied under the influence of magnetic fields [58]. In the gel the dispersed particles align uniaxially before application of a magnetic field. The increase in shear modulus has been detected in magnetic fields of the order of 10 kA/m. As seen from Eq. 39, ΔG is proportional to C or H^2 when χ is independent of H. Figure 12 indicates the effect of the volume fraction of the particles on $\Delta G'$ and $\Delta G''$. The MR effect is strongly reflected in the storage modulus. The experimental results are in good accordance with the simple model of the point dipole. As the value of χ in an iron particle was kept constant in the intensity range of 0 to 80 kA/m, the relation between ΔG and H was investigated in this intensity range. $\Delta G'$ and $\Delta G''$ have been related to H^2.

An MR hydrogel of a crosslinked PVA network and Fe_3O_4 particles dispersed at random was also synthesized [59]. The viscosity of the gel in a magnetic field increased with time. It has been suggested that the dispersed Fe_3O_4 particles aggregate and form a cluster. The effect of the volume fraction of the particles on the viscosity has also been discussed. The authors have pointed out that when the volume fraction is about 15%, the ratio of the viscosities with and

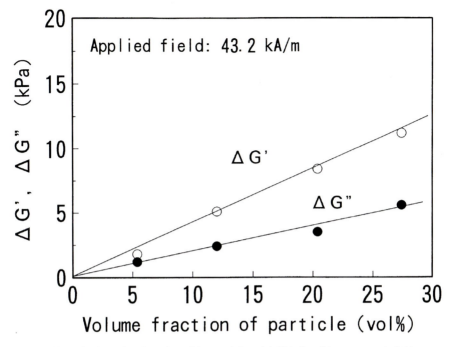

Fig. 12. Effect of volume fraction of particles on $\Delta G'$ and $\Delta G''$ induced by a magnetic field

without a magnetic field attains a maximum. However, this phenomenon has not been interpreted.

In summary, the dynamic viscoelasticity of ER or MR gels in electric or magnetic fields has been investigated. The nature of the ER or MR effects in polymer gels may be roughly described using the model of the point dipole. We have now established that the rigidity and the damping property of the composites can be controlled by external signals.

5.2 Polymer Processing

The formation of migration paths of dispersed particles as shown in Fig. 9 suggests a new approach for the modulation of polymer blends and morphology. It is expected that ER or MR effects show not only smartness but also unique anisotropy; therefore, many workers are studying these effects from the point of view of polymer processing.

Moriya et al. first demonstrated that anisotropic structure in blends can be generated by applying ac electric fields to solvent-free blends of PEO and PS [60]. A similar morphology transformation induced by electric fields has been detected in PEO/PS/cyclohexane systems [61–63]. During solvent evaporation, elongated PEO phases are first observed, and pearl chains of spheres are then formed. It is believed that the elongated domains are initially formed more easily and are stabilized against breakup as a result of a lowering of interfacial tension between PEO and PS.

The alignment of a lamellar microstructure by electric fields has been reported [64]. The electric fields were applied across a melt of a PS-PMMA block copolymer and were maintained throughout cooling down to below the glass transition point. SAXS studies show persuasive evidence that the microstructure was aligned by an electric field.

Electric field-induced fibrolysis of suspended micron-sized powders such as ceramics, ferroelectric metals, and semiconductors has been investigated in a variety of uncured thermoset polymers [65, 66]. The resulting material consists of a quasi 1–3 connectivity pattern due to the formation of fibrils. For a given thermoset polymer, the degree to which these fibrils form is dependent upon both the magnitude and frequency of the applied field. Pattern formation in colloidal dispersions of $BaTiO_3$ (100 nm diameter) under electric fields has been studied. A dilute dispersion of $BaTiO_3$ in castor oil has been injected into clear castor oil through a pinhole. When an electric field of 2000 V/cm is applied near the pinhole, the dilute suspension is deformed into a prolate ellipsoid [67].

6 Application

6.1 Membranes and Drug Delivery Systems

Stimuli-responsive hydrogels may swell or deswell sharply with relatively small changes in environmental conditions. The stimuli-responsive hydrogels have promising potential to achieve intelligent membrane separation or drug delivery systems because they may be utilized as molecular devices for automatically regulating them with a function of sensing. The permeability-controlled membranes are applicable to time-variable separation or transport of a solute, as well as small scale analytical systems. The drug delivery systems using polymer gels are being explored to overcome the disadvantages of conventional dosage forms such as tolerance problems.

The transport of ^3H_2O and ^{14}C-sucrose through a collagen membrane in an electric field has been studied by Eisenberg and Grodzinsky [68]. It has been demonstrated that the application of an electric field can change the permeability of the membrane. The field has selectively changed the membrane's permeability to ^{14}C-sucrose. In contrast, permeability to ^3H_2O is minimally affected.

An electrically activated membrane which reversibly expands and contracts in pore size has been designed by other researchers [8]. The gel membrane used was a PAMPS gel. The water permeability through the membrane was controlled by electrical signals.

Concerning drug delivery, electrically erodible polymer gels for controlled release of drugs have been prepared, and a measured release rate of insulin has been observed under electrical stimulus [69]. A suspension of zinc insulin in a mixed solution of poly(ethyloxazoline) and PMAA was formed into a gel by decreasing the pH of the suspension. The obtained complex gel with 0.5 wt% of insulin was attached to a woven platinum wire cathode which was 1 cm away from the anode and immersed in 0.9% saline solution. When a stepped function of electrical current of 5 mA was applied to the insulin-loaded gel matrix, insulin was released in a stepwise manner up to a release of 70%. The insulin rate measured was 0.10 mg/h.

A fast-acting smart polymer gel for drug delivery systems has recently been designed [19]. The polymer gel is a secretory granule matrix obtained from mice, consisting of a negatively charged heparin sulfate proteoglycan network. The granule matrix condenses in the presence of divalent cations and decondenses in the presence of monovalent cations. The granule matrix of 5 μm diameter was attached to one electrode in a solution, and an electric voltage was applied. At zero or positive potentials, the granule matrix was refractile and condensed. In contrast, the matrix was transparent and swollen at negative potentials. The deformation took place within milliseconds of application of an electric field of the order of 1 kV/cm (8 V per 5 μm).

6.2 Artificial Muscle

Systems that develop contractile forces are very intriguing as analogues of physiological muscles. The idea for gel muscles was based upon the work of Katchalsky and Kuhn. They have prepared polyelectrolyte films or fibers which become elongated or contracted in response to a change in pH of the surrounding solution, and have estimated the induced force and response time. The contraction of gel fibers is also achieved by electric fields. Use of electric fields has the merit that the signals are easily controlled.

The model of electric field-controlled artificial muscles has been described in 1972 [5]. Fragala et al. fabricated an electrically activated artificial muscle system which uses a weakly acidic contractile polymer gel sensitive to pH changes. The pH changes are produced through electrodialysis of a solution. The response of the muscle as a function of pH, solution concentration, compartment size, certain cations, and gel fabrication has been studied. The relative change in length was about 10%, and the tensile force was 1 g/0.0025 cm^2 under an applied electric field of 1.8 V and 10 mA/cm^2. It took 10 min for the gel to shrink.

Tensile forces electrically induced in a charged-collagen membrane which supports a gradient in neutral salt concentration have also been measured [7]. The observed force density of oriented collagen fiber was 1 N/cm^2 within 5 min on application of an electric field with a current of 0.12 mA/cm^2. This value was much smaller than the force density induced by an increase in NaCl concentration.

The electrocontractile response of a composite of a PVA-PAA gel doped with PPy has been investigated by a group in the University of Pisa [70]. The composite strip (initial length 3.05 cm, thickness 0.016 cm) in a 0.05 N NaCl solution has acted as both an electrode and an actuator. When dc voltages up to 2 V were applied, PPy doped in the gel underwent a redox reaction to change the pH inside the gel, and the change in size was observed within 30 min. The measured length variation was about 2%.

A new composite PVA hydrogel for an artificial muscle has been prepared by a freezing and thawing method [71]. The gel contained PAA and PAAm.HCl (poly-allylamine hydrochloride). The electrocontractile behavior of the composite gel in various solutions was studied. A large stroke and better controllability have been detected in a 10 mM NaOH/7 mM Ba(OH)$_2$ system.

6.3 Biomimetic Actuator

PVA-PAA gel is a physically crosslinked gel of PVA chains entangled with PAA chains. PVA-PAA gel has good mechanical properties, particularly rubber elasticity. It shows an electric field-associated bending deformation as observed in polyelectrolyte gel. The attractive point of the bending of PVA-PAA gel is that the deformation occurs at relatively high speeds. For example, a gel of 70 mm length, 7 mm width, and 7 mm thickness bends semicircularly within 30 s on application of 10 V/cm. My laboratory team has made a mechanical hand composed of four smart PVA-PAA gel fingers [72, 73]. The mechanical hand can pick up and hold a fragile quail egg (9 grams) without breaking it in an Na$_2$CO$_3$ solution in response to an electrical signal (Fig. 13a).

The strain in electric field-associated bending of a PVA-PAA gel is given by the equation $\varepsilon = 6DY/L^2$ (see Eq. 21). The strain depends on the electric power applied to the gel. Thus, the deflection increases as the thickness becomes small even if the electric power remains constant. The PVA-PAA gel rod of 1 mm diameter bends semicircularly within 1 s under both dc and ac excitation. An artificial fish with a PVA-PAA gel tail 0.7 mm thick has been designed, and it has been demonstrated that the fish swims forward at a velocity of 2 cm/sec as the gel flaps back and forth under sinusoidally varied electric fields (Fig. 13b). This prototype of a biomimetic actuator shows that translational motion may be produced using bending deformation [74].

An electric field-induced bending of gel makes a worm-like motion feasible [18]. A weakly crosslinked PAMPS gel in a surfactant solution bends toward the anode under dc electric fields. Both ends of the gel are placed on front and rear hooks and are then hung on a plastic ratchet bar. When a varying electric field of 10 V/cm and 0.5 Hz is applied, the gel moves forward in the solution with a bending motion at a velocity of 25 cm/min.

The three biomimetic actuator models mentioned above are driven in a solution. The next target of the advanced model is an actuator electrically driven in air. A mechanical hand composed of two smart gel fingers working in

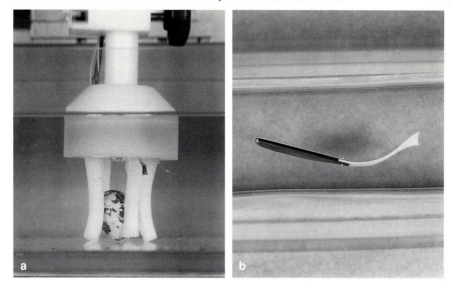

Fig. 13a, b. Biomimetic actuators using electric field-responsive gels: **a** robot hand having four smart gel fingers which can hold a quail egg, and **b** artificial fish with a tail of gel film which can swim under ac electric fields

air and using the hybrid gel has been described in Sect. 3.2 [29]. Two platinum wires of 50 μm diameter as electrodes were provided in the PVA gel films of the hybrid gel. The PVA film of the hybrid gel plays the role of a solution, which provides water. The mechanical hand with the two gel fingers can bend the fingers inward simultaneously and catch a piece of paper of 2 g in air. When the polarity of the applied voltage is changed, the fingers release the paper.

6.4 Vibration Isolator

There are many sounds and vibrations in automobiles such as booming noises, harshness, arm chattering, or microshakes due to an imbalance of the engine. Many researchers have recently attempted to reduce these sounds and vibrations actively in order to achieve comfortable driving. The smart ER or MR gels in Sect. 5 are applicable to actively controlled vibration isolators which can change damping properties in response to an external signal. A suspension mount in which an ER gel and a rubber support are laminated together has been proposed [75]. The designed mount can absorb vibration energy by application of electric signals. A vibration bushing using an ER gel has also been designed [76]. However, the active damping properties of these vibration isolators have not yet been reported.

7 Conclusions

Advanced polymer gels are now being produced with stimuli-responsive properties and have become "smarter". In this article, smart gels associated with
electric fields are described. Although an electric field is only one among several
possible stimuli, its use may well offer the most promising route to practical
applications. With regard to the electric field-induced deformation of gels, the
invention of bending is exploitable and may open a new door in gel-based
technology. The feasibility of a large deflection in the bending increases the
feasibility of biomimetic actuators. The observation of ER or MR effects in
polymer gels leads one to envision a new immiscible polymer blend that can
vary its elastic modulus actively in an electric field. Such a material could lead to
surprising developments in the fields of robotics and automobiles.

A host of scientists in universities and industrial laboratories are finding that
by adding side chains to the crosslinked chains or otherwise altering their
structure they can control when and how the materials respond. It is therefore to
be expected that they will design new smart gels and ideas for gel-using systems
in the future.

Acknowledgements I would like to thank Professor M. Doi of Nagoya University for stimulating
discussions. I am also very grateful to Y. Hirose and Dr. T. Kurauchi of Toyota Central Research
and Development Laboratories, who helped me to fabricate biomimetic actuators.

8 References

1. DeRossi D, Kajiwara K, Osada Y, Yamauchi A (1991) Polymer gels. Plenum Press, New York
2. Tanaka T (1978) Phys Rev Lett 40: 820–823
3. Tanaka T (1981) Sci Am 244: 110–123
4. Hamlen RP, Kent CE, Shafer SN (1965) Nature 206: 1149–1150
5. Fragala A, Enos J, LaConti A, Boyack J (1972) Electrochim Acta 17: 1507–1522
6. Grodzinsky AJ, Shoenfeld NA (1977) Polymer 18: 435–443
7. Shoenfeld NA, Grodzinsky AJ (1980) Biopolymers 19: 241–262
8. Osada Y, Hasebe M (1985) Chem Lett 1285–1287
9. Kishi R, Osada Y (1989) J Chem Soc Faraday Trans 85: 655–662
10. Tanaka T, Nishio I, Sun S-T, Ueno-Nishio S (1982) Science 218: 467–469
11. Hirotsu S (1985) Jpn J Appl Phys 24: 396–398
12. Shiga T, Kurauchi T (1985) Polymer Preprints Japan 34: 509
13. Shiga T, Kurauchi T (1990) J Appl Polym Sci 39: 2305–2320
14. Doi M, Matsumoto M, Hirose Y (1992) Macromolecules 25: 5504–5511
15. Irie M (1986) Macromolecules 19: 2476–2477
16. Osada Y, Okuzaki O, Hori H (1992) Nature 355: 242–244
17. Shinohara H, Aizawa M (1989) Jpn J Polym Sci Technol 46: 703–708
18. Tanaka H, Shiga T, Hirose Y, Okada A, Kurauchi T (1993) Jpn J Polym Sci Technol 50: 963–967
19. Nanavati C, Fernandez JM (1993) Science 259: 963–965
20. Flory PJ (1953) Principles of Polymer Chemistry. Cornell Univ Press, Ithaca New York
21. Moore WJ (1977) Physical Chemistry, 4th edn. Tokyo Kagaki Doujin Tokyo
22. Hirose Y, Giannetti G, Marquardt J, Tanaka T (1992) J Phys Soc Jpn 61: 4085–4097
23. Shiga T, Hirose Y, Okada A, Kurauchi T (1992) J Appl Polym Sci 46: 635–640
24. Nanbu M (1983) Polymer Appl 32: 523–530

25. Hyon SH, Cha WI, Ikada Y (1989) Jpn J Polym Sci Technol 46: 673–680
26. Shiga T, Fukumori K, Hirose Y, Okada A, Kurauchi T (1994) J Appl Polym Sci 32: 85–90
27. Suzuki M (1991) Int J Jpn Soc Prec Eng 25: 169–174
28. Shiga T, Hirose Y, Okada A, Kurauchi T (1992) J Appl Polym Sci 44: 249–252
29. Shiga T, Hirose Y, Okada A, Kurauchi T (1993) J Intel Mater Systems Structures 4: 553–557
30. Shiga T, Hirose Y, Okada A, Kurauchi T (1993) J Appl Polym Sci 47: 113–119
31. Shiga T, Hirose Y, Okada A, Kurauchi T (1994) J Maters Sci 29: 5715–5718
32. Hurty WC, Rubinstein MF (1968) Dynamics of structures. Prentice Hall, New York
33. Nakazawa H, Nagaya Z, Katoh H (1973) Zairyo rikigaku. Sangyo Tosho, Tokyo
34. Adriani PM, Gast AP (1988) Phys Fluids 31: 2757–2768
35. Deinega YF, Vinogradov GV (1984) Rheologica Acta 23: 636–651
36. Hill JC, van Steenkiste TH (1991) J Appl Phys 70: 1207–1211
37. Halsey TC, Martin JE, Adolf D (1991) Phys Rev Lett 68: 1519–1522
38. Winslow WM (1947) US Patent 2, 417, 850
39. Klass DL, Martinek TW (1967) J Appl Phys 38: 67–74
40. Uejima H (1972) Jpn J Appl Phys 11: 319–325
41. Stangroom JE (1983) Phys Technol 14: 290–292
42. See H, Tamura H, Doi M (1993) J Phys D Appl Phys 26: 746–752
43. Block H, Kelly JP, Watson T (1991) Spec Publ R Soc Chem 87: 151–173
44. Klingenberg DJ, van Swol F, Zukosky CF (1989) J Chem Phys 91: 7888–7895
45. Whittle M (1990) J Non-Newtonian Fluid Mechs 37: 233–263
46. See H, Doi M (1991) J Phys Soc Jpn 60: 2778–2782
47. Takimoto Z (1992) J Rheol Soc Jpn 20: 95–100
48. Pohl HA (1978) Dielectrophoresis. Cambridge Univ Press, Cambridge, UK
49. Sprecher AF, Chen Y, Conrad H (1989) Proc 2nd Inter Conference on ER Fluids, 82–89
50. Shiga T, Ohta T, Hirose Y, Okada A, Kurauchi T (1991) Jpn J Polym Sci Tecnol 48: 47–51
51. Kumagai N (1987) Dengikigaku Kisoron. Ohm Book Company, Tokyo
52. Yamamoto S, Matsuoka T, Takahashi H, Kurauchi T (1992) J Rheol Soc Jpn 2056–60
53. Plonsey R, Collin RE (1961) Principles and application of electromagnetic fields. McGraw-Hill, New York
54. Shiga T, Ohta T, Hirose Y, Okada A, Kurauchi T (1993) J Maters Sci 28: 1293–1299
55. Shiga T, Okada A, Kurauchi T (1993) Macromolecules 26: 6958–6963
56. Shiga T, Okada A, Kurauchi T (1995) J Maters Sci Lett 14: 514–515
57. Ishino Y, Hukuyama Y, Saitoh T (1992) Japanese Laid-open Patent Publication 211931–1992
58. Shiga T, Okada A, Kurauchi T (1995) J Appl Polym Sci 58: 787–792
59. Fujikura Y, Kondou S, Hoshino H (1996) Polymer Preprints Jpn 45: 784
60. Moriya S, Adachi K, Kotaka K (1985) Polym Commun 26: 235–239
61. Venugopal G, Krause S, Wnek GE (1989) J Polym Sci Part C Polym Lett 27: 497–501.
62. Serpico JM, Wnek GE, Krause S, Smith TW, Luca DJ, van Laeken (1991) Macromolecules 24: 6879–6881
63. Serpico JM, Wnek GE, Krause S, Smith TW, Luca DJ, van Laeken (1992) Macromolecules 25: 6373–6374
64. Amundson K, Helfand E, Davis DD, Quan X, Patel SS (1991) Macromolecules 24: 6546–6548
65. Randall CA, Miyazaki S, More KL, Bhalla AS, Newnham RE (1993) J Mater Res 8: 899–905
66. Bowen CP, Bhalla AS, Newnham RE, Randall (1994) J Mater Res 9: 781–788
67. Trau M, Sankaran S, Saville DA, Aksay IA (1995) Nature 374: 437–439
68. Eisenberg SR, Grodzinsky AJ (1984) J Membrane Sci 19: 173–194
69. Kwon IC, Bae YH, Kim SW (1991) Nature 354: 291–293
70. DeRossi D, Chiarelli P, Buzzigoli G, Domenici C, Lazzeri L (1986) Trans Am Soc Artif Inter Organs 17: 157–162
71. Suzuki M (1989) Proc Int Workshop on Intell Maters 307–312
72. Kurauchi T, Shiga T, Hirose Y, Okada A (1990) MRS Polym Symp Procs 171: 389–393
73. Shiga T, Hirose Y, Okada A, Kurauchi T (1989) Jpn J Polym Sci Technol 46: 709–713
74. Shiga T, Hirose Y, Okada A, Kurauchi T (1989) Proc 1st Jpn Inter SAMPE Symp. 659–663
75. Kurokawa Y, Mouri N, Nakai K, Hirose Y, Shiga T (1993) Japanese Laid-open Patent Publication 79534–1993
76. Maeno T (1992) Japanese Laid-open Patent Publication 107334–1992

Editor: A. Abe
Received: September 1996

Rheology of Polymers Near Liquid-Solid Transitions

Horst Henning Winter and Marian Mours
University of Massachusetts, Department of Chemical Engineering and
Department of Polymer Science and Engineering, Amherst, MA 01003, USA

Polymeric materials near the liquid-solid transition (LST) exhibit a very distinct relaxation pattern. The reference point for analyzing these patterns is the instant of LST at which relaxation becomes self-similar over wide ranges of the relaxation time. The universality of this transition and its consequences have been explored extensively during the past decade. This study will present an overview of rheological implications inherent in liquid-solid transitions of polymers. The LST can be most reliably detected in a dynamic mechanical experiment in which the frequency independence of the loss tangent marks the LST. A wide variety of rheological observations of materials in the vicinity of an LST are discussed with respect to their universality. It is shown that polymer chemistry, molecular weight, stoichiometry, temperature, inhomogeneities, etc. greatly influence the material behavior near the LST. However, the characteristic self-similar relaxation is shown by all investigated materials, independent of the nature of the LST (e.g., both, physically and chemically crosslinking polymers). Several theories predict chemical and rheological properties in the vicinity of an LST. They are briefly discussed and compared with experimental results. A variety of applications for polymers near LST are presented that either already exist or can be envisioned. The self-similar relaxation behavior which results in a power law relaxation spectrum and modulus is not restricted to materials near LST. Different classes of polymers are described that also show power law relaxation behavior. What makes the self-similar relaxation specific for materials at LST is its occurrence at long times with the longest relaxation time diverging to infinity.

Advances in Polymer Science, Vol. 134
© Springer-Verlag Berlin Heidelberg 1997

List of Abbreviations

BSW   Baumgärtel, Schausberger, Winter (spectrum for linear flexible chains of uniform length) [61]
CW   Chambon-Winter (spectrum for critical gel)
FS   Flory-Stockmayer
LCP   liquid crystalline polymer
LST   liquid/solid transition
ODT   order/disorder transition
PBD   polybutadiene
PDMS   polydimethylsiloxane
UV   ultraviolet

List of Symbols

A	constant
a_T	horizontal shift factor
α_-	critical exponent for longest relaxation time before LST
α_+	critical exponent for longest relaxation time after LST
b_T	vertical shift factor
β	scaling exponent for gel fraction
$\mathbf{C}(t; t')$	Cauchy strain tensor
$\mathbf{C}^{-1}(t; t')$	Finger strain tensor
c	concentration
d	space dimension
d_f	fractal dimension
$\bar{d}_f$	fractal dimension with excluded volume screening
δ	phase angle
δ_c	phase angle at LST
E/R	activation energy divided by universal gas constant
F_{ij}	fragmentation kernel, probability of cluster of size i + j to break up into cluster of size i and cluster of size j
f	functionality, frequency
$\bar{f}_2$	average number of crosslinking sites along a chain
f_g	gel fraction
G	relaxation modulus
G_0	plateau modulus of fully crosslinked material
G_e	equilibrium modulus
G^*	complex modulus
G'	storage modulus
G''	loss modulus

G'_c	storage modulus at LST
G''_c	loss modulus at LST
$\bar{g}_2$	average number of crosslinking sites along a chain
g	measured property in definition of mutation number
Γ	Gamma function
γ	shear strain, critical exponent for molecular weight
γ_0	step shear strain, shear strain amplitude
$\dot{\gamma}$	shear rate
$\dot{\gamma}_0$	constant shear rate
H	relaxation spectrum
H_0	front factor of power law spectrum
h	Heaviside step function
η'	real part of complex viscosity, $\eta' = G''/\omega$
η''	imaginary part of complex viscosity, $\eta'' = G'/\omega$
η_0	zero-shear viscosity
J	creep compliance
J^0_e	equilibrium compliance
J'	storage compliance
J''	loss compliance
K_{ij}	reaction kernel, probability of cluster of size i to react with cluster of size j
κ	scaling exponent of rates of change of dynamic moduli
Λ	exponent for certain reaction kernel K_{ij}, $\Lambda = \mu + \nu$
λ	relaxation time
λ_0	lower bound of CW relaxation spectrum, characteristic relaxation time of liquid state, characteristic material time
λ_b	lifetime of physical bond
λ_{char}	characteristic relaxation time
λ_l	lower cutoff relaxation time of power law spectrum
λ_{max}	longest relaxation time
λ_{pg}	lifetime of physical cluster
λ_u	upper cutoff relaxation time of power law spectrum
M	cluster mass
M_e	entanglement molecular weight
M_{GAUSS}	molecular weight above which chains behave Gaussian
M_{max}	molecular weight of largest cluster
M_n	number average molecular weight
M_w	weight average molecular weight
m	power law exponent (for spectrum with positive exponent)
μ	exponent for homogeneous reaction kernel K_{ij}
N	number of bonds in a molecular cluster
$N(M)$	cluster mass distribution
N_1	first normal stress difference
N_g	gel number
N_{max}	maximum number of bonds in a molecular cluster

N_{mu}	mutation number
n	relaxation exponent
n_{Af}	number of molecules of A of functionality f
n_{Bg}	number of molecules of B of functionality g
n_f	number of molecules of functionality f
ν	critical exponent for typical cluster size
p	measure of connectivity, e.g. extent of reaction in case of chemical gelation
p_c	critical extent of reaction
p_A	extent of reaction of species A
p_B	extent of reaction of species B
π	osmotic pressure
R	radius of gyration
R_{char}	typical cluster size
r	molar ratio
r_1	lower molar ratio
r_u	upper molar ratio
ρ	cluster density, mass density
S	gel stiffness
s	critical exponent for zero-shear viscosity
σ	critical exponent for maximum molecular weight
T	temperature
ΔT	degree of supercooling
T_0	reference temperature
T_c	temperature at critical point
T_g	glass transition temperature
T_m	melt temperature
t	time
Δt	sampling time
t'	time (integration variable)
t''	time between t' and t
t_1	creep time
t_p	process time
τ	critical exponent for cluster mass distribution, stress
τ	stress tensor
$\dot{\tau}$	rate of change of shear stress
τ_0	constant applied shear stress in creep experiment
τ_1	stress under static load at infinite time
$\tau_{11}-\tau_{22}$	first normal stress difference
$\tau_{21} = \tau_{12}$	shear stress
υ	exponent for homogeneous reaction kernel K_{ij}
ω	angular frequency
ξ	correlation length
ψ_1	coefficient of first normal stress difference
z	critical exponent for equilibrium modulus

1 Patterns of Relaxation Near the Liquid-Solid Transition

Polymeric materials relax with a broad spectrum of relaxation modes. Magnitude and shape of the spectrum reflect the material structure in some complicated way. The longer modes belong to the motion of entire molecules or of large chain segments while the shorter modes characterize small scale details of the molecules. Extra long relaxation modes arise from large scale structures which some polymers are able to form due to phase separation or associations on the molecular or particulate level. The formation of such extensive clusters is the origin of many liquid-solid transitions. Most intriguing is the behavior near such transitions, when molecular motions slow down while they correlate with motions of other molecules over longer and longer distances. The relaxation modes are not independent any more, but they are somehow coupled over a wide range of time scales. This leads to a universal pattern of the relaxation time spectrum at liquid-solid transitions. The universality of the rheological behavior and its consequences have been explored extensively during the past decade, and this study will attempt to give an overview of the current state of the field.

There are many reasons for studying the liquid-solid transition (LST). The physicist might be interested in gelation as a critical phenomenon. The LST of polymers is also technically important since it occurs in nearly all of the common fabrication processes. Examples are injection molding of semi-crystalline polymers (where the surface quality of the finished parts may be affected by gelation) and processing of crosslinking polymers. The instant of LST has to be known for the design and operation of such polymer processing. The polymer processing engineer may like to anticipate the instant of gelation, often for the mere reason of avoiding or postponing it. Beyond that, processing near the gel point promises interesting texture development for ultimate material properties. The materials scientist might like to know the possible range of material properties close to the gel point. Conservation of the material state near an LST has the potential for novel properties which combine liquid and solid characteristics. Industrial applications are just beginning to explore such advantageous properties in adhesives, super absorbers, dampers, sealants, membranes, toner matrices, catalyst supports, etc. Gels are good adhesives since they combine the surface wetting property of liquids with the cohesive strength of solids. Strong adhesion and damping properties recommend gels as binder material in composite materials. Widespread technical applications have not yet materialized, since, until recently, it has been difficult to measure and control the LST. This has changed, and, as a consequence, one is able now to control processing near LST or to manufacture gels with reproducible properties.

The chemical gel point defines the instant of LST of *chemically crosslinking* polymers. Before the crosslinking polymer has reached its gel point it consists of a distribution of finite clusters. It is called a 'sol' since it is soluble in good solvents. Beyond the gel point, it is called a 'gel'. The gel is an infinitely large macromolecule which only can swell but not dissolve in a solvent, even if low molecular weight molecules (sol fraction) are still extractable from the gel. We will later borrow this terminology ('gel point', 'sol', 'gel') from chemical gelation and apply it to a wide range of materials which share rheological properties with chemically crosslinking systems. These are the *physical gels*, which are able to form extensive molecular or particulate clusters by a variety of different mechanisms. Examples are partially crystalline polymers, liquid crystalline polymers at their nematic-to-smectic transition, micro-phase separating block copolymers, and suspensions and emulsions at the percolation limit. Emphasis in this study will be on the rheological behavior, without trying to discuss the various 'crosslinking' mechanisms.

The independent variable, p, of the solidification process differs from material to material. It is a measure of connectivity (see Fig. 1), which requires restatement for each type of LST. An exception is chemical gelation for which the extent of crosslinking reaction, p, is defined and directly measurable as the ratio of the number of chemical bonds to the total number of possible bonds ($0 < p < 1$, without ever reaching unity), i.e. p is the bond probability. At the critical extent of reaction, $p \rightarrow p_c$, the molecular weight of the largest molecule diverges to infinity and the molecular weight distribution spreads infinitely broad ($M_w/M_n \rightarrow \infty$), i.e. molecular sizes range from the smallest unreacted oligomer to the infinite cluster. This defines the gel point [1, 2]. The value of p_c is not universal but depends on the details of the evolving structure.

The polymer at the gel point is in a critical state [3], and the name *critical gel* [4] is appropriate for distinguishing polymers at the gel point from the various materials which commonly are called *gels*. The critical gel has no intrinsic size scale except for the size of its oligomeric building block, and molecular motions are correlated over large distances. The combination of liquid and solid

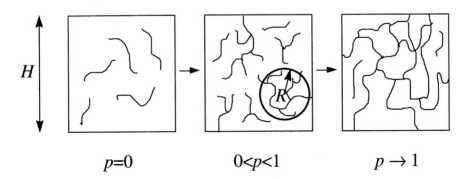

Fig. 1. Schematic of cluster growth during crosslinking. At $p = 0$, only the monomer is present. With increasing crosslinking index, p, the connectivity increases and the molecular clusters (radius R) grow in size. In the solid state, the network spans the entire sample, $2R \rightarrow H$

behavior at the gel point requires unusual simplicity and regularity in the relaxation pattern.

Very important are materials in the vicinity of the gel point. For such *nearly critical gels*, $p - p_c$ is a measure of the distance from the gel point, and all properties can be expanded in powers of $|p - p_c|$. This is permitted within the critical region at small distance at both sides of the gel point [3]. Material properties of nearly critical gels are still governed by the simplicity of the critical state. This changes at increased $|p - p_c|$, where the behavior loses its simplicity. It will be interesting to study the rheological properties which gradually break free from that pattern as the distance from the gel point increases.

Material properties at a critical point were believed to be independent of the structural details of the materials. Such universality has yet to be confirmed for gelation. In fact, experiments show that the dynamic mechanical properties of a polymer are intimately related to its structural characteristics and forming conditions. A direct relation between structure and relaxation behavior of critical gels is still unknown since their structure has yet evaded detailed investigation. Most structural information relies on extrapolation onto the LST.

1.1 Rheological Observations of a Liquid-Solid Transition

The transition strongly affects the molecular mobility, which leads to large changes in rheology. For a direct observation of the relaxation pattern, one may, for instance, impose a small step shear strain γ_0 on samples near LST while measuring the shear stress response $\tau_{12}(t)$ as a function of time. The result is the shear stress relaxation function $G(t) = \tau_{12}(t)/\gamma_0$, also called relaxation modulus. Since the concept of a relaxation modulus applies to liquids as well as to solids, it is well suited for describing the LST.

Figure 2 shows a typical evolution of $G(t, p)$ near the LST of a crosslinking polymer. The x axis shows the time of crosslinking reaction which corresponds to an extent of reaction, p. For each of the curves in Fig. 2, p is kept constant. The crosslinking reaction was stopped at discrete values of p, which increased from sample to sample.

In samples with early stages of crosslinking (lower curves in Fig. 2), stress can relax quickly. As more and more chemical bonds are added, the relaxation process lasts longer and longer, i.e. $G(t)$ stretches out further and further. The downward curvature becomes less and less pronounced until a straight line ('power law') is reached at the critical point.

Exactly at the LST, the material behaves not as a liquid any more and not yet as a solid. The relaxation modes are not independent of each other but are coupled. The coupling is expressed by a power law distribution of relaxation modes [5–7]

$$G(t) = St^{-n} \quad \text{for } \lambda_0 < t < \infty \tag{1-1}$$

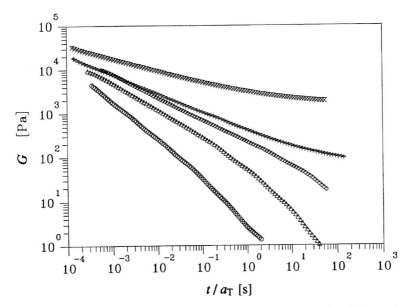

Fig. 2. Relaxation modulus $G(t)$ of a set of polydimethylsiloxane samples with increasing extent of crosslinking plotted against time of crosslinking. The linear PDMS chains ($M_n \approx 10\,000$, polydispersity index ≈ 2) were endlinked with a four-functional silane crosslinker catalyzed by a platinum compound. Samples with different extent of reaction were prepared by poisoning the reaction at different times. The actual extent of reaction was not determined. Two of the samples are clearly before the gel point (LST) and two beyond. The third sample is very close to the gel point. Data of Chambon and Winter [5] evaluated by Baumgärtel and Winter [8]

S being the gel stiffness as indicated by the straight line in the log/log plot, Fig. 2. This marks the intermediate state between curving down and curving to the right, and we assume that the power law behavior extends to infinite times. The power law may be explained by the hypothesis that one probes self-similar regions of the critical gel by varying the times of observations [4]. The upper cut-off is infinite since the longest relaxation time diverges to infinity at the LST. Parameters S, n, and the lower cross-over, λ_0, depend on the material structure at the transition.

Beyond the LST, $p > p_c$, the material is a solid. The solid state manifests itself in a finite value of the relaxation modulus at long times, the so-called equilibrium modulus

$$G_e = \lim_{t \to \infty} G(t). \tag{1-2}$$

Stresses cannot relax completely any more. The upper curves in Fig. 2 show this curving to the right, where at long times an equilibrium stress level will eventually be reached. More data at longer times would be required in order to clearly identify the value of G_e. However, G_e can be estimated from the curves, and it can be seen that G_e is zero at the gel point and grows with the extent of reaction, p.

Rubbery materials beyond the gel point have been studied extensively. A long time ago, Thirion and Chasset [9] recognized that the relaxation pattern of a stress τ under static conditions can be approximated by the superposition of a power law region and a constant limiting stress τ_1 at infinite time:

$$\tau = \tau_1 \left(1 + \left(\frac{t}{\lambda_0} \right)^{-n} \right) \tag{1-3}$$

where λ_0 is a material-dependent time constant. They found very low values for the slope n in the power law region, $0.12 < n < 0.17$.

1.2 Relaxation Time Spectrum

The linear viscoelastic behavior of liquid and solid materials in general is often defined by the relaxation time spectrum $H(\lambda)$ [10], which will be abbreviated as 'spectrum' in the following. The transient part of the relaxation modulus as used above is the Laplace transform of the relaxation time spectrum $H(\lambda)$

$$G(t) = G_e + \int_0^{\lambda_{max}} H(\lambda) e^{-t/\lambda} \frac{d\lambda}{\lambda}. \tag{1-4}$$

The spectrum is a non-negative function [11] which exists in the range of relaxation times $0 < \lambda \leqslant \lambda_{max}$. An important material property is the longest relaxation time, λ_{max}, beyond which the spectrum is equal to zero; $H(\lambda) = 0$ for $\lambda > \lambda_{max}$. The spectrum cannot be measured directly. However, many methods have been proposed to somehow extract $H(\lambda)$ from linear viscoelastic material functions as measured in the appropriate experiments. A comprehensive review of some of those methods was recently presented by Orbey and Dealy [12].

We assume that the spectrum $H(\lambda)$ gradually evolves as the material undergoes transition. There exists a spectrum for the material directly at the transition point, the critical gel. Its characteristic features are twofold: a longest relaxation time (upper limit of the integral) that diverges, $\lambda_{max} \to \infty$, and a power law distribution with a negative exponent, $-n$. Both properties are expressed in the self-similar CW spectrum which Chambon and Winter [6, 7] found when analyzing chemical gelation experiments (Fig. 3):

$$H(\lambda) = \frac{S}{\Gamma(n)} \lambda^{-n} \quad \text{for } \lambda_0 < \lambda < \infty \tag{1-5}$$

where $\Gamma(n)$ is the gamma function. Stress relaxation is the same at all scales of observation for such 'self-similar' or 'scale invariant' behavior. It is interesting to note that the critical gel does not have a characteristic time constant, which is a rather unusual property for a viscoelastic material.

The relaxation exponent n is restricted to values between 0 and 1. The case of $n = 0$ corresponds to the limiting behavior of a Hookean solid (the relaxation

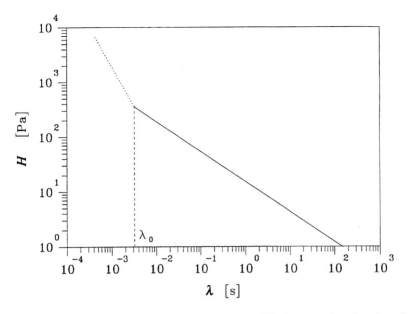

Fig. 3. Schematic of Chambon-Winter gel spectrum. The longest relaxation time diverges to infinity. The relaxation time λ_0 marks the crossover to the short-time behavior, which depends on the material. The depicted case corresponds to a low-molecular-weight precursor (crossover to glass transition region)

modulus is a constant). The restriction of n to values less than unity is necessary to assure a diverging zero-shear viscosity at the gel point.

The self-similar spectrum is not valid at short times, $\lambda < \lambda_0$, where the details of chemical structure become important (glass transition, entanglements, etc.). The cross-over to the glass transition at short times is typical for all polymeric materials, for both liquids and solids. The critical gel is no exception in that respect. λ_0 could be used as a characteristic time in the CW spectrum since it somehow characterizes the molecular building block of the critical gel; however, it has no direct relation to the LST. At times shorter than λ_0, the LST has no immediate effect on the rheology. Indirect effects might be seen as a shift in the glass transition, for instance, but these will not be studied here.

1.3 Divergence of Longest Relaxation Time

In the close vicinity of the gel point, $|p_c - p| \ll 1$, the longest relaxation time diverges in a power law on both sides of the gel point (Fig. 4)

$$\lambda_{max} \propto \begin{cases} (p_c - p)^{-\alpha_-} & \text{for sol, } p < p_c \\ (p - p_c)^{-\alpha_+} & \text{for gel, } p > p_c. \end{cases} \tag{1-6}$$

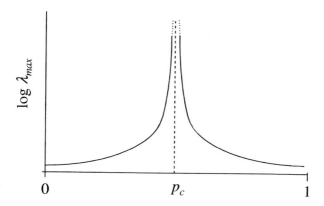

Fig. 4. Schematic of the divergence of the longest relaxation time as the liquid-solid transition is approached from either side

α_- and α_+ are the critical exponents for the sol and the gel. In the sol, λ_{max} belongs to the largest cluster. The largest cluster reaches infinite size at the gel point, but it still can relax, and the corresponding λ_{max} has become infinitely large. Beyond the gel point, the relaxable components (for chemical gelation this would be the sol fraction, unattached chain ends, long loops, etc.) gradually incorporate into the permanent network, and λ_{max} decays again.

With increasing distance from the gel point, the simplicity of the critical state will be lost gradually. However, there is a region near the gel point in which the spectrum still is very closely related to the spectrum at the gel point itself, $H(\lambda, p_c)$. The most important difference is the finite longest relaxation time which cuts off the spectrum. Specific cut-off functions have been proposed by Martin et al. [13] for the spectrum and by Martin et al. [13], Friedrich et al. [14], and Adolf and Martin [15] for the relaxation function $G(t, p_c)$. Sufficiently close to the gel point, $|p - p_c| \ll 1$, the specific cut-off function of the spectrum is of minor importance. The problem becomes interesting further away from the gel point. More experimental data are needed for testing these relations.

It was a most interesting discovery that not only the longest relaxation time diverges at LST, but that all the shorter relaxation modes show a very distinct pattern. The longest mode escapes the measurement near LST while the spectrum of the shorter modes is still accessible. The intent of this study is to explore the occurrence of this relaxation time spectrum in a broad range of solidifying materials and in a time or frequency window which is as wide as possible. The properties of the self-similar CW spectrum, Eq. 1–5, will be mapped out in Sect. 3. The behavior at LST then will serve as a reference state for the analysis of rheological phenomena in the vicinity of LST. This will set the stage for reviewing experimental data from several laboratories. Observations on chemical gelation will guide the analysis of various types of LST. The possibility will be suggested that there exists a universal framework for many LSTs of different origin. In that spirit, the terms 'liquid-solid transition' (LST) and 'gel point' will be used synonymously.

1.4 Interrelation Between Critical Exponents

Steady shear flow properties are sensitive indicators of the approaching gel point for the liquid near LST, $p < p_c$. The zero shear viscosity η_0 and equilibrium modulus G_e grow with power laws [16]

$$\eta_0 \propto (p_c - p)^{-s} \qquad \text{for sol, } p < p_c \tag{1-7}$$

$$G_e \propto (p - p_c)^z \qquad \text{for gel, } p > p_c \tag{1-8}$$

having critical exponents, s and z. The viscosity of the sol increases due to the diverging cluster size. The equilibrium modulus of the gel gradually builds up since an increasing fraction of the molecules join, and thereby strengthen the sample spanning permanent network (Fig. 5).

As a result, we find for sols that the divergence of the above zero shear viscosity η_0 and of two other linear viscoelastic material functions, first normal stress coefficient ψ_1 and equilibrium compliance J_e^0, depends on the divergence of λ_{max} [17]

$$\lambda_{max}(p) \propto \eta_0^{1/(1-n)} \propto \psi_1^{1/(2-n)} \propto J_e^{0^{1/n}} \qquad \text{for sol, } p < p_c. \tag{1-9}$$

Only the value of the relaxation exponent is needed. The critical exponent α_- of the longest relaxation time (compare Eqs. 1-6 and 1-7) is therefore on an equal footing with the critical exponent of the viscosity:

$$s = (1 - n)\alpha. \tag{1-10}$$

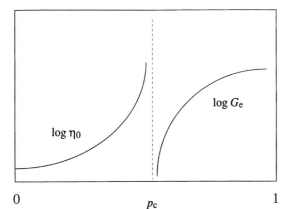

0   p_c 1

Fig. 5. Schematic of the divergence of zero-shear viscosity, η_0, and equilibrium modulus, G_e. The LST is marked by p_c

For the relaxation of the solid near the gel point, the critical gel may serve as a reference state. The long time asymptote of $G(t)$ of the nearly critical gel, the equilibrium modulus G_e, intersects the $G(t) = St^{-n}$ of the critical gel at a characteristic time (Fig. 6) which we will define as the longest relaxation time of the nearly critical gel [18]

$$G_e = (St^{-n})_{t = \lambda_{max}} \rightarrow \lambda_{max} = \left(\frac{G_e(p)}{S} \right)^{-1/n} \quad \text{for gel, } p > p_c. \quad (1\text{-}11)$$

It obeys the typical characteristics, namely the divergence to infinity as G_e goes to zero (gel point) and the approach of a zero value as G_e becomes large. Again, only the relaxation exponent n is needed for relating the divergence of λ_{max} with that of G_e; compare Eqs. 1-8 and 1-11:

$$z = n\alpha_+ \quad \text{for gel, } p > p_c. \quad (1\text{-}12)$$

The exponents α_- and α_+ depend not only on the relaxation exponent n, but also on the dynamic exponents s and z for the steady shear viscosity of the sol and the equilibrium modulus of the gel.

The analysis may be simplified by postulating symmetry of the diverging λ_{max} on both sides of the gel point [18]. A power law exponent (see Eq. 1-6) which is the same on both sides,

$$\alpha = \alpha_- = \alpha_+ \quad (1\text{-}13)$$

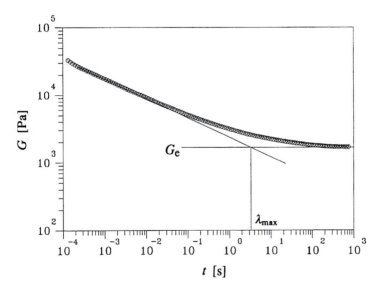

Fig. 6. Evaluation of the longest relaxation time for a sample beyond the gel point, $p > p_c$: intersect of horizontal line for G_e with the power law of the critical gel, St^{-n}

leads to the interesting relations between critical exponents [13, 18, 19]

$$n = z/(z + s). \tag{1-14}$$

$$\alpha = s + z, \tag{1-15}$$

$$s = (1 - n)\alpha; \ z = n\alpha. \tag{1-16}$$

Only two of the exponents (α and n, for instance) are sufficient to describe the rheology of nearly critical gels. The front factor is more difficult to estimate, but it most likely differs on both sides.

  These relations will be useful for testing theories, since, except for the symmetry hypothesis, no specific assumptions were introduced in the derivation. Theory might give an answer about the validity of the above symmetry hypothesis. In fact, the theory of Goldbart and Goldenfeld [20] that is based on statistical mechanics yields Eq. 1-14. However, there does not seem to be an easy way of proving or disproving this hypothesis at this time. The wide range of values for the relaxation exponent, $0 < n < 1$, lets us expect that the dynamic exponents s and z are non-universal as well.

2 Theory of Gelation

This study is mostly concerned with experimental aspects, especially since a quantitative prediction of the self-similar spectrum (value of critical exponent and prefactor) from first principles seems to be still lacking, although several theories predict the evolution of cluster growth during gelation. Excellent reviews of theory have been given by Stauffer et al. [3, 16] and Vilgis [21]. We refer to these for a deeper study and only highlight several of the theoretical predictions in the following.

2.1 Branching Theories

Branching models are based on multifunctional molecules of different types between which covalent bonds are formed to yield a network structure. One of the multifunctional molecules is required to carry at least three functional groups, while the other one can have two functional groups. The overall extent of reaction, p, equals the a priori probability that any given functional group has condensed. The earliest of these branching theories was developed by Flory [1, 22] and Stockmayer [2]. Using combinatorial approaches, they derived an expression for the molecular weight distribution, and subsequently the critical extent of reaction, p_c, at which the molecular weight diverges, $M_w \to \infty$ (gel point). Their approach includes several simplifying assumptions which are usually not valid in real systems, i.e. (1) the reactivities of all functional groups of the same type are equal and independent of each other, (2) no intramolecular reactions between functional groups on the same cluster ('loop formation') are allowed, (3) the crosslinks are randomly formed between any pair of functional groups that can form a bond, and (4) point-like monomers are assumed (no steric hindrance and excluded volume effects). More advanced branching models were developed later. The two most widely used of these are the so-called recursive theory [23, 24] and the cascade theory [25–28]. These later models can deal approximately with nonidealities such as cyclization and long-range substitution effects. All branching theories are mean field theories and yield the same simple expression for the critical extent of reaction (for the same chemical model) depending on the geometry of the network. Special cases are:

Case 1. Homopolymerization of similar f-functional molecules:

$$p_c = \frac{1}{\bar{f}_2 - 1} \tag{2-1}$$

The same relation is found for the end-linking of molecules of low functionality ($f = 3$ or 4) and for the vulcanization of long molecular chains. The second-moment average number of cross-linking sites along the chain, $\bar{f}_2$, is defined as

$$\bar{f}_2 = \frac{\sum_f f^2 n_f}{\sum_f f n_f} \tag{2-2}$$

with n_f = number of molecules of functionality f.

Case 2. Cross-linking of f-functional molecules A_f with g-functional molecules B_g, which are mixed at a molar ratio $r = \sum_f f n_{Af} / \sum_g g n_{Bg}$

$$p_{A,c} = \frac{1}{\sqrt{r(\bar{f}_2 - 1)(\bar{g}_2 - 1)}} \tag{2-3}$$

with $p_B = r p_A$. The stoichiometric ratio of a sample must be chosen between a lower and upper critical value

$$r_1 = \frac{1}{(\bar{f}_2 - 1)(\bar{g}_2 - 1)}; \quad r_u = \frac{1}{r_1} \tag{2-4}$$

otherwise the reaction stops before reaching the gel point. The relations in Eq. 2-4 follow from Eq. 2-3 when considering species A_f or species B_g fully reacted, respectively.

Experimental results of p_c, r_1, and r_u were found to agree reasonably well with these predictions despite the inherent assumptions [29–31].

2.2 Percolation Theory

Percolation theory describes [32] the random growth of molecular clusters on a d-dimensional lattice. It was suggested to possibly give a better description of gelation than the classical statistical methods (which in fact are equivalent to percolation on a Bethe lattice or Caley tree, Fig. 7a) since the mean-field assumptions (unlimited mobility and accessibility of all groups) are avoided [16, 33]. In contrast, immobility of all clusters is implied, which is unrealistic because of the translational diffusion of small clusters. An important fundamental feature of percolation is the existence of a critical value p_c of p (bond formation probability in random bond percolation) beyond which the probability of finding a percolating cluster, i.e. a cluster which spans the whole sample, is non-zero.

In random bond percolation, which is most widely used to describe gelation, monomers, occupy sites of a periodic lattice. The network formation is simulated by the formation of bonds (with a certain probability, p) between nearest neighbors of lattice sites, Fig. 7b. Since these bonds are randomly placed between the lattice nodes, intramolecular reactions are allowed. Other types of percolation are, for example, random site percolation (sites on a regular lattice are randomly occupied with a probability p) or 'random random' percolation (also known as continuum percolation: the sites do not form a periodic lattice but are distributed randomly throughout the percolation space). While the

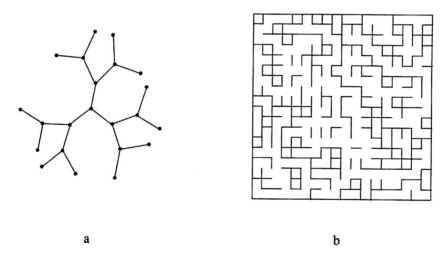

a   b

Fig. 7. Comparison of **a** the structure of the Bethe lattice with a functionality of 3 (only part of the system is shown) and **b** a two-dimensional square lattice [16]. For the Bethe lattice, each possible bond is shown as a line connecting two monomers. In FS theory an actual bond of these possible bonds is formed with probability p. For the square lattice, each bond that has been formed is shown as a short line connecting two monomers, while the monomers are not shown

random site percolation is not directly relevant to gelation [16], continuum percolation is of particular value, since in real systems the cluster-forming molecules are not distributed regularly in space.

In general, percolation is one of the principal tools to analyze disordered media. It has been used extensively to study, for example, random electrical networks, diffusion in disordered media, or phase transitions. Percolation models usually require approximate solution methods such as Monte Carlo simulations, series expansions, and phenomenological renormalization [16]. While some exact results are known (for the Bethe lattice, for instance), they are very rare because of the complexity of the problem. Monte Carlo simulations are very versatile but lack the accuracy of the other methods. The above solution methods were employed in determining the critical exponents given in the following section.

2.3 Scaling Near the LST

All theories yield unique scaling relationships for molecular (e.g. mean cluster size, size distribution) and bulk properties (e.g. equilibrium modulus) near the critical point, but critical exponent values and relations between different critical exponents are different. This scaling is common for material behavior near any critical point, i.e. the polymeric material near the gel point exhibits a behavior analogous, for example, to a fluid near its vapor-liquid critical point. For the critical gel, weight average molecular weight M_w, typical cluster size R_{char}, and gel fraction f_g scale similarly with $|p - p_c|$ as the inverse of the derivative of osmotic pressure with respect to concentration $(\partial \pi / \partial c)^{-1}$, correlation length ξ, and concentration fluctuations Δc, respectively, scale with $T - T_c$ in case of a fluid at the vapor-liquid critical point [3, 34]. The following scaling relationships for these static properties are commonly found in the literature [16, 35]:

$$M_w \propto |p - p_c|^{-\gamma} \qquad p < p_c \tag{2-5}$$

$$R_{char} \propto |p - p_c|^{-\nu} \qquad p < p_c \tag{2-6}$$

$$f_g \propto |p - p_c|^{\beta} \qquad p > p_c \tag{2-7}$$

The cluster mass distribution at the gel point scales with the molecular weight of those clusters

$$N(M) \propto M^{-\tau} \qquad p = p_c \tag{2-8}$$

To describe the cluster mass distribution in the vicinity of the gel point, a cut-off function $f(M/M_{max})$ is introduced [36] (Fig. 8)

$$N(M) \propto M^{-\tau} f\left(\frac{M}{M_{max}}\right) \tag{2-9}$$

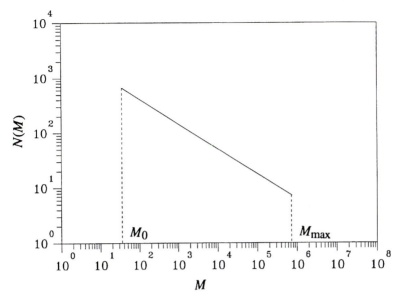

Fig. 8. Power law molecular weight distribution in the vicinity of the LST. M_0 is the molecular weight of the smallest precursor molecule, and M_{max} the molecular weight of the largest cluster present in the polymer

with

$$M_{max} \propto |p - p_c|^{-1/\sigma} \tag{2-10}$$

Percolation theory predictions for these critical exponents are $\gamma = 1.76$, $\nu = 0.89$, $\beta = 0.39$, $\tau = 2.2$ and $\sigma = 0.46$. The Flory-Stockmayer theory also predicts this scaling behavior near the gel point, with exponents $\gamma = 1$, $\nu = 0.5$, $\beta = 1$, $\tau = 2.5$, and $\sigma = 0.5$ [3, 37].

Colby et al. [35] proposed an interesting experimental approach to measure the static exponents. They noticed that it is hard to accurately measure the chemical extent of reaction, p, and thus eliminated this variable (more precisely the distance from the gel point $|p - p_c|$) from the scaling relations. For example combining Eqs. 2-5 and 2-6 yields the following relation between the weight average molecular weight, M_w, and the characteristic radius, R_{char}:

$$M_w \propto R_{char}^{\gamma/\nu} \tag{2-11}$$

Similar relations between different scaling exponents were also developed by Stauffer [37] by combining two of the scaling relations at a time to eliminate $|p - p_c|$.

Besides the static scaling relations, scaling of dynamic properties such as viscosity η and equilibrium modulus G_e [16, 34], see Eqs. 1-7 and 1-8, is also predicted. The equilibrium modulus can be extrapolated from dynamic experiments, but it actually is a static property [38].

The scaling of the relaxation modulus $G(t)$ with time (Eq. 1-1) at the LST was first detected experimentally [5–7]. Subsequently, dynamic scaling based on percolation theory used the relation between diffusion coefficient and longest relaxation time of a single cluster to calculate a relaxation time spectrum for the sum of all clusters [39]. This resulted in the same scaling relation for $G(t)$ with an exponent n following Eq. 1-14.

It is interesting to note here that the cluster mass distribution and the relaxation modulus $G(t)$ at the LST scale with cluster mass and with time, respectively, while all other variables (dynamic and static) scale with the distance from p_c in the vicinity of the gel point.

The classical theory predicts values for the dynamic exponents of $s = 0$ and $z = 3$. Since $s = 0$, the viscosity diverges at most logarithmically at the gel point. Using Eq. 1-14, a relaxation exponent of $n = 1$ can be attributed to classical theory [34]. Dynamic scaling based on percolation theory [34, 40] does not yield unique results for the dynamic exponents as it does for the static exponents. Several models can be found that result in different values for n, s and z. These models use either Rouse and Zimm limits of hydrodynamic interactions or Electrical Network analogies. The following values were reported [34, 39]: (Rouse, no hydrodynamic interactions) $n = 0.66$, $s = 1.35$, and $z = 2.7$, (Zimm, hydrodynamic interactions accounted for) $n = 1$, $s = 0$, and $z = 2.7$, and (Electrical Network) $n = 0.71$, $s = 0.75$ and $z = 1.94$.

De Gennes [41] predicted that percolation theory should hold for crosslinking of small molecule precursors. However, he argued that for vulcanizing polymers (high M_w), only a very narrow regime near the gel point exists for which percolation is valid, i.e. these polymers should exhibit more mean field-like behavior.

2.4 Critical Gel as Fractal Structure

Based on the fractal behavior of the critical gel, which expresses itself in the self-similar relaxation, several different relationships between the critical exponent n and the fractal dimension d_f have been proposed recently. The fractal dimension d_f of the polymer cluster is commonly defined by [16, 42]

$$R \propto M^{1/d_f} \tag{2-12}$$

where R is the radius of gyration. Assuming hyperscaling d_f can be related to the critical exponents by (not valid for mean field theories) [16, 34]

$$d_f = d - \frac{\beta}{v} \tag{2-13}$$

This results in a value of $d_f = 2.5$ for bond percolation on a 3-dimensional lattice. The fractal dimension of the Bethe lattice (Flory-Stockmayer theory) is

$d_f = 4$ [16], which leads to a physical discrepancy, since any value higher than 3 (dimensionality of the sample) results in the cluster density increasing with cluster size ($\rho \propto R^{d_f - d}$). For $d = 6$ the hyperscaling assumption and mean field theories are compatible, i.e. Eq. 2-13 gives the correct fractal dimension for the classical theory.

Muthukumar and Winter [42] investigated the behavior of monodisperse polymeric fractals following Rouse chain dynamics, i.e. Gaussian chains (excluded volume fully screened) with fully screened hydrodynamic interactions. They predicted that n and $\bar{d}_f$ (the fractal dimension of the polymer if the excluded volume effect is fully screened) are related by

$$n = \frac{\bar{d}_f}{\bar{d}_f + 2} \tag{2-14}$$

Hess et al. [43] extended this approach to monodisperse chains with excluded volume effects (swollen clusters). They realized that, although the linkage process can be described by percolation, bond percolation does not give a correct picture of crosslinking between long chains because these chains are flexible, whereas bond percolation theory is based on stiff bonds. Thus, even though the connectivity of the critical gel may be prescribed by bond percolation theory, the dynamic properties of the object are drastically affected by the replacement of rigid bonds by flexible chains. Their investigation resulted in the same functional dependence of n and d_f (the fractal dimension when excluded volume effects are taken into account):

$$n = \frac{d_f}{d_f + 2} \tag{2-15}$$

where d_f and $\bar{d}_f$ are related by

$$\bar{d}_f = \frac{2d_f}{d + 2 - 2d_f} \tag{2-16}$$

Muthukumar [44] further investigated the effects of polydispersity, which are important for crosslinking systems. He used a hyperscaling relation from percolation theory to obtain his results. If the excluded volume is not screened, n is related to d_f by

$$n = \frac{d_f}{d_f + 2} \tag{2-17}$$

In the case of full screening of excluded volume he obtained

$$n = \frac{d}{\bar{d}_f + 2} = \frac{d(d + 2 - 2d_f)}{2(d + 2 - d_f)} \tag{2-18}$$

Especially in the latter case, a small change in the fractal dimension can lead to a significant change in n, and he therefore concluded that n can take values between 0 and 1 (for d_f ranging from 2.5 to 1.25, see Fig. 9).

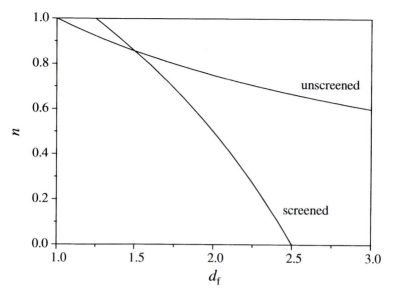

Fig. 9. Relation between relaxation exponent n and fractal dimension d_f for a three-dimensional network. In case of complete screening of excluded volume, values of $0 < n < 1$ are possible if d_f is chosen between 1.25 and 2.5

If only partial screening is present, the fractal dimension takes a value somewhere between d_f and $\bar{d}_f$. According to this model, a crosslinker deficiency, which leads to a more open structure and therefore a lower value of d_f, increases the value of n. Dilution of the precursor with a non-reactive species has the same effect on the relaxation exponent.

2.5 The Notion of Topology

Goldbart and Goldenfeld [20] challenged the notion that gelation could be described in terms of the purely geometrical description of percolation theory. They developed a theory based on statistical mechanics arguments. The cross-linking system cannot be uniquely specified by the positions of crosslinks only, as is done in percolation theory. Topology needs to be considered as well. They argue that the condition for the liquid-solid transition is a sufficiently complex topology rather than a sufficient degree of connectivity, in contrast to percolation, which does not take the contribution of trapped entanglements into account. They define the solid state in terms of a non-vanishing shear modulus (as $t \to \infty$) and in terms of breaking of the translational invariance of the Hamiltonian. This implies that rigidity of the solid state is due to the preference of atoms to localize close to certain neighbors in order to minimize the free energy of the system rather than due to long range forces, i.e. rigidity is a consequence of thermodynamics. Percolation models do not address the question of how rigidity emerges. The new model is a microscopic theory of the liquid-solid transition based on a physical model of flexible chains and solvent. The transition to the solid state as the crosslink density increases beyond the critical density is a continuous one, and hence is a second order transition. This is rather unusual for liquid-solid transitions, which are usually first order. The second order transition is accompanied by scale invariance, and therefore implies scaling behavior of the shear modulus as detected by Chambon and Winter [5], but it does so only at the critical point. Goldenfeld and Goldbart [20] developed relations between scaling exponents z, s, and n for the equilibrium modulus G_e, the zero-shear viscosity η_0, and the complex modulus $G(t)$, respectively, and predict Eq. 1-14, $n = z/(s + z)$, from some general arguments. The exponent of the equilibrium modulus is predicted to equal the scaling exponent for the correlation length ξ. However, the theory does not predict any values for these scaling exponents, which makes comparison with experimental data difficult.

2.6 Kinetic Theory (Smoluchowski Equation)

All models described up to here belong to the class of equilibrium theories. They have the advantage of providing structural information on the material during the liquid-solid transition. Kinetic theories based on Smoluchowski's coagulation equation [45] have recently been applied more and more to describe the kinetics of gelation. The Smoluchowski equation describes the time evolution of the cluster size distribution $N(k)$:

$$\frac{dN(k)}{dt} = \frac{1}{2} \sum_{i+j=k} K_{ij} N(i) N(j) - N(k) \sum_{j=1}^{\infty} K_{kj} N(j) \tag{2-19}$$

$N(k)$ denotes the number of clusters of size k (i.e. k-mers), and K_{ij} is the reaction kernel that gives the probability of a cluster of size i reacting with one of size j. The first sum accounts for coalescence of clusters of size i and $(k - i)$ to give a cluster of size k, while the second sum accounts for the loss of clusters of size k due to binary collisions with other clusters. The Smoluchowski equation is able to describe and distinguish between gelling and non-gelling systems. In the former, the mean cluster size diverges as t approaches the gel point t_c; in the latter it keeps increasing with time. Although the equation was originally developed only for irreversible coagulation, it can be easily extended to reversible coagulation by adding fragmentation kernels F_{ij} to describe the unimolecular fragmentation process [46]:

$$\frac{dN(k)}{dt} = \frac{1}{2} \sum_{i+j=k} \{K_{ij} N(i) N(j) - F_{ij} N(k)\}$$

$$- \sum_{j=1}^{\infty} \{K_{kj} N(k) N(j) - F_{kj} N(k + j)\} \tag{2-20}$$

where F_{ij} describes the probability of a $(i + j)$ -mer to break up and form and i-mer and a j-mer. Both Equations 2-19 and 2-20 constitute infinite sets of coupled non-linear equations which have to be solved for a given initial cluster size distribution $N(k, t_0)$.

The difficulty is now to determine the functional form of the reaction and fragmentation kernels, K_{ij} and F_{ij}. The specific form is determined by the coagulation mechanism. For gelation processes, usually only the reaction kernel K_{ij} is considered, i.e. the process is viewed as being irreversible. A variety of kernels for coagulation processes can be found in the literature [47]. Most of these kernels are homogeneous functions of i and j, at least for large i and j. Ernst [47] and Van Dongen and Ernst [48] used two exponents μ and v to describe the i and j dependence of $K_{ij} = K(i, j)$:

$$K(ai, aj) = a^\Lambda K(i, j) = a^\Lambda K(j, i) \qquad (2\text{-}21a)$$

$$K(i, j) \propto i^\mu j^v \qquad (2\text{-}21b)$$

with $j \gg 1$ and $\lambda = \mu + v$. Two physical restrictions exist on the exponents μ and v, because the reaction rate cannot increase faster than the cluster size: $\Lambda = \mu + v \leqslant 2$ and $v \leqslant 1$. Λ characterizes the reaction rate of two large inter-penetrable clusters, i.e. $K(j, j) \propto j^{\mu + v}$, and v describes the reaction of a large cluster with a very small cluster, i.e. $K(1, j) \propto j^v$. Furthermore, Λ also decides whether the Smoluchowski equation describes aggregation or gelation, i.e. the formation of an infinite cluster in finite time only occurs if $\Lambda > 1$.

Three growth classes are usually distinguished. These are class I with $\mu > 0$, class II with $\mu = 0$ and class III with $\mu < 0$. In the case of class I, growth interactions between two large clusters are dominant. Class I growth can describe both aggregating and gelling systems. For gelation, $v \leqslant 1 < \Lambda \leqslant 2$. The cluster distribution decays as $N(k) \propto k^{-\tau}$ with $\tau = \frac{1}{2}(\Lambda + 3) > 2$. Also, near the gel point, the weight average mass defined as $M_w = \sum k^2 N(k) / \sum k N(k)$ diverges as $M_w \propto |t - t_c|^{-\gamma}$ with $\gamma = (3 - \Lambda)/(\Lambda - 1)$, the typical cluster mass (the z-average mass, $M_z = \sum k^3 N(k) / \sum k^2 N(k)$) diverges as $M_z \propto |t - t_c|^{-1/\sigma}$ with $\sigma = (\Lambda - 1)/2$, and the gel fraction diverges as $f_g \propto |t - t_c|^\beta$ with $\beta = 1$ [49]. These scaling equations correspond to Eqs. 2-8, 2-5, 2-10, and 2-7, respectively, as presented earlier in this Section. If one uses the Flory-Stockmayer gelation theory, the reaction kernel equals $K_{ij} = ij$, since all sites on a cluster are assumed to be equally reactive [50]. This is a typical example of a homogeneous kernel which gives class I growth. The exponents are therefore $\mu = 1$ and $v = 1$, resulting in $\Lambda = 2$. This results in scaling exponents $\tau = 2.5$, $\gamma = 1$, $\sigma = 0.5$, and $\beta = 1$, which are also predicted by the FS-theory (see section 2.1.).

In class II growth, the large-large (class I) and small-large (class III) interactions are equal. Since $\mu = 0$, it follows that $v = \Lambda$. Because of the restriction on $v (v \leqslant 1)$ and the requirement of $\Lambda > 1$ for gelation this class can only describe non-gelling growth. Interactions between small and large clusters govern class III growth. From $\mu < 0$, it follows that $v > \Lambda$, i.e. class III growth is defined by $\Lambda < v \leqslant 1$. Like class II growth, it can only describe aggregation.

2.7 Computer Simulations

Alternatively, Leung and Eichinger [51] proposed a computer simulation approach which does not assume any lattice as the classical and percolation theory. Their simulations are more realistic than lattice percolation, since spatially closer groups form bonds first and more distant groups at later stages of network formation. However, the implicitly introduced diffusion control is somewhat obscure. The effects of intramolecular reactions were more realistically quantified, and the results agree quite well with experimental observations [52, 53].

3 Viscoelastic Properties at and around the Liquid-Solid Transition

3.1 Linear Viscoelasticity of Liquids and Solids

The time-dependent rheological behavior of liquids and solids in general is described by the classical framework of linear viscoelasticity [10, 54]. The stress tensor τ may be expressed in terms of the relaxation modulus $G(t)$ and the strain history:

$$\tau(t) = \int_{-\infty}^{t} \frac{\partial G(t - t')}{\partial t'} \mathbf{C}^{-1}(t; t') \, dt' \qquad (3\text{-}1)$$

or, alternatively:

$$\tau(t) = -\int_{-\infty}^{t} G(t - t') \frac{\partial \mathbf{C}^{-1}(t; t')}{\partial t'} \, dt'. \qquad (3\text{-}2)$$

The relaxation modulus is often expressed with the relaxation time spectrum, Eq. 1-4:

$$\tau(t) = G_e \mathbf{C}^{-1}(t; t_0) + \int_{-\infty}^{t} \int_{0}^{\lambda_{max}} H(\lambda) e^{-(t-t')/\lambda} \frac{d\lambda}{\lambda^2} \mathbf{C}^{-1}(t; t') \, dt' \qquad (3\text{-}3)$$

Here we describe the strain history with the Finger strain tensor $\mathbf{C}^{-1}(t;t')$ as proposed by Lodge [55] in his rubber-like liquid theory. This equation was found to describe the stress in deforming polymer melts as long as the strains are small (second strain invariant below about 3 [56]). The permanent contribution $G_e \mathbf{C}^{-1}(t; t_0)$ has to be added for a linear viscoelastic solid only. $\mathbf{C}^{-1}(t; t_0)$ is the strain between the stress free state t_0 and the instantaneous state t. Other strain measures or a combination of strain tensors, as discussed in detail by Larson [57], might also be appropriate and will be considered in future studies. A combination of Finger $\mathbf{C}^{-1}(t; t')$ and Cauchy $\mathbf{C}(t; t')$ strain tensors is known to express the finite second normal stress difference in shear, for instance.

The stress τ is more difficult to model for a material which is deforming while solidifying. A crosslinking polymer, for instance, constantly introduces new crosslinks even during relaxation processes. Molecules or segments of molecules gradually lose their mobility, since the crosslinking locks them into their relative positions with neighboring molecules. As a starting point, we consider materials for which the crosslinking reaction has been stopped or slowed down so severely that they may be treated as chemically stable during a rheological experiment. This quasi-stability will be explained further in Sect. 6.2.

3.2 Viscoelastic Material Functions of Critical Gels

We expect that the classical framework of linear viscoelasticity also applies at the gel point. The relaxation spectrum for the critical gel is known and can be inserted into Eq. 3-3. The resulting constitutive equation will be explored in a separate section (Sect. 4). Here we are mostly concerned about the material parameters which govern the wide variety of critical gels.

The linear viscoelastic behavior of the critical gel, as defined in Eqs. 1-1 and 1-5, depends on two material parameters, the relaxation exponent n and the front factor S. Depending on their values, the critical gel is more soft or more stiff. The relaxation exponent strongly depends on molecular and structural details which affect the development of long range connectivity. These are molecular weight of the precursor, stoichiometric ratio, amount and molecular weight of inert diluent in the material, and bulkiness of the crosslinks (filler effect). The relaxation exponent does not have a universal value, as one might expect for a property at a critical point.[1] The critical gel is generally very soft and fragile when the relaxation exponent is large, $n \to 1$, and the front factor S is small. Stiff critical gels have a small n value ($n \to 0$) and a large S. For many systems, the front factor is not independent, but depends on the relaxation exponent

$$S = G_0 \lambda_0^n, \tag{3-4}$$

where G_0 and λ_0 are the plateau modulus of the fully crosslinked material and the characteristic time of the precursor molecule (building block of the gel), respectively [58, 59]. This ensures a soft gel for $n \to 1$ and a stiff gel for $n \to 0$. Figure 10 shows data of S and n measured by Izuka et al. [59] on polycaprolactone critical gels with different stoichiometric ratios. The dashed line connects the modulus of the fully crosslinked material and the zero-shear viscosity of the precursor, which is in the order of $G_0 \lambda_0$.

The gel stiffness, S, was also found to depend on the molecular weight of the polymer precursor. For end-linking PDMS, S decreases with increasing

---

[1] Our first two critical gels had an exponent value of $n \approx 0.5$, which made us believe initially that this would be the only possible value

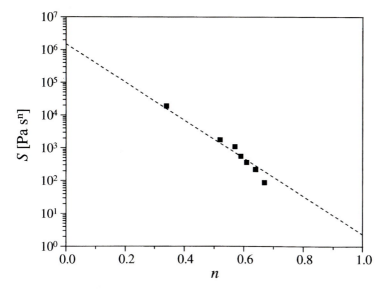

Fig. 10. Experimental values of the gel stiffness S plotted against the relaxation exponent n for crosslinked polycaprolactone at different stoichiometric ratios [59]. The *dashed line* connects the equilibrium modulus of the fully crosslinked material (on left axis) and the zero shear viscosity of the precursor (on right axis)

M_w because of an increase in strand length between crosslinks [7]. In contrast, vulcanizing polybutadienes of high molecular weight ($M_w >$ entanglement molecular weight) show a relaxation exponent of about or somewhat below 0.5 and an increase in gel stiffness with increasing precursor molecular weight. For this study, De Rosa and Winter [60] crosslinked polybutadienes (PBD) with long linear molecules of (nearly) uniform molecular weight and measured the relaxation time spectrum at increasing extents of reaction (Fig. 11 depicts the critical gel spectrum for the PBD with a molecular weight of 44 000). The precursor molecular weight was chosen to be ten or more times the entanglement molecular weight. The precursor's relaxation follows the BSW spectrum [61, 62]. Near the gel point, the plateau modulus of the entanglement region (intermediate time scales) is surprisingly little affected by the crosslinking (Fig. 12). Only at higher extents of crosslinking, beyond the gel point, does the plateau modulus start to increase significantly. A scaling relation between S and M was found to be valid for these materials:

$$S \propto M^{3.4n} \tag{3-5}$$

where 3.4 is the well known scaling exponent of the viscosity-molecular weight relation.

For describing the observed molecular weight effects in chemical gels, we propose to distinguish three regions based on the precursor molecular

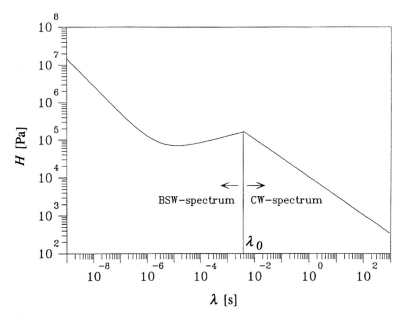

Fig. 11. Schematic of relaxation time spectrum of the critical gel of PBD 44 ($M_w = 44\ 000$). The entanglement and glass transition is governed by the precursor's BSW-spectrum, while the CW spectrum describes the longer modes due to the crosslinking [60]. λ_0 denotes the longest relaxation time of PBD44 before crosslinking

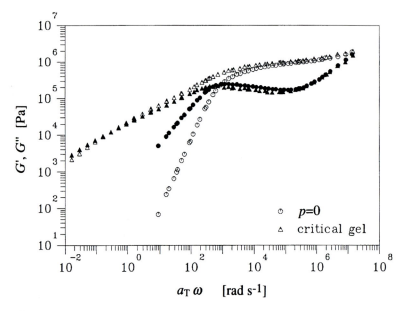

Fig. 12. Dynamic moduli master curves of PBD 44 precursor ($p = 0$) and PBD 44 critical gel [60]. The entanglement and glass transition regime is hardly affected by the crosslinking. *Open symbols* correspond to G', *filled ones* to G''

weight M:

1. Low molecular weight M: $\qquad M < M_{GAUSS}$: $\qquad n \approx 0.7–0.8$
2. Intermediate molecular weight: $\quad M_{GAUSS} < M < M_e$: $n \approx 0.4–0.7$
3. High molecular weight: $\qquad\qquad M > M_e$: $\qquad n \approx 0.4–0.5$.

M_{GAUSS} is the molecular weight above which chains show Gaussian behavior. Very short chains, $M < M_{GAUSS}$, associate into molecular clusters with non-Gaussian building blocks, and the resulting critical gel is very stiff. Extensive data have been reported in the literature on these systems [38, 39, 63, 64]. Intermediate molecular weight precursors, $M_{GAUSS} < M < M_e$, already give much lower relaxation exponents [5–7, 18, 58, 65]. The lowering of the relaxation exponent has been attributed to screening of excluded volume and hydrodynamic interaction [44].

Deficiency of cross-linker molecules (off-balancing of stoichoimetry) was found to increase the relaxation exponent value [7, 65, 66]. The gel becomes more 'lossy', and stress relaxation is accelerated. Adding of a non-reacting low molecular weight solvent also increases the relaxation exponent [58, 65], even in physical gels [67]. Both effects have been attributed to screening [44, 65].

On the other hand, 'bulky' crosslinks as developed during the crystallization of polymer melts (no solvent) lower the relaxation exponent. The lowest values of n which we have been able to generate so far occurred with physical gels in which the crosslinks consisted of large crystalline regions [68, 69].

This regular pattern in the relaxation exponent has been recognized for a wide range of chemically and physically gelling systems. The full range of gel properties should be explored further and should be utilized technically. The molecular or structural origin of these variations is not yet known to the extent where quantitative predictions could be derived from first principles. From a practical point of view, it is advantageous that the relaxation exponent is non-universal, since it allows us to prepare nearly critical gels with a wide range of properties as needed for specific applications.

3.3 Viscoelastic Material Functions Near LST

The simplest expression incorporating the basic features of self-similarity and cut-off for nearly critical gels has the spectrum of the critical gel altered by a cut-off at the longest time [19]:

$$
H(\lambda, p) = \begin{cases} \dfrac{S}{\Gamma(n)} \lambda^{-n} & \text{for } \lambda_0 < \lambda < \lambda_{max}(p) \\ 0 & \text{for } \lambda > \lambda_{max}(p) \end{cases},
\tag{3-6}
$$

The same form of self-similar spectrum will be applied to the sol and the transient part of the gel. The consequences of this most simple spectrum will be explored in the following. Introducing Eq. 3-6 into the equation for the relaxation modulus, Eq. 1-4, gives

$$G(t) = G_e(p) + \frac{S}{\Gamma(n)} \int_0^{\lambda_{max}(p)} \lambda^{-n} e^{-t/\lambda} \frac{d\lambda}{\lambda} \tag{3-7}$$

The diverging longest relaxation time, Eq. 1-6, sets the upper limit of the integral. The solid (gel) contribution is represented by G_e. The crossover to any specific short-time behavior for $\lambda < \lambda_0$ is neglected here, since we are mostly concerned with the long-time behavior.

We can also calculate other viscoelastic properties in the limit of low shear rate (linear viscoelastic limit) near the LST. The above simple spectrum can be integrated to obtain the zero shear viscosity η_0, the first normal stress coefficient ψ_1 at vanishing shear rate, and the equilibrium compliance J_e^0:

$$\eta_0(p) = \int_0^{\lambda_{max}} H(\lambda) \, d\lambda = \frac{S\lambda_{max}^{1-n}}{(1-n)\Gamma(n)}, \tag{3-8}$$

$$\psi_1(p) = 2 \int_0^{\lambda_{max}} H(\lambda)\lambda \, d\lambda = \frac{2S\lambda_{max}^{2-n}}{(2-n)\Gamma(n)}, \tag{3-9}$$

$$J_e^0 = \frac{\psi_1}{2\eta_0^2} = \frac{\Gamma(n)(n-1)^2}{S} \frac{2-n}{\lambda_{max}^n}. \tag{3-10}$$

This most simple model for the relaxation time spectrum of materials near the liquid-solid transition is good for relating critical exponents (see Eq. 1-9), but it cannot be considered quantitatively correct. A detailed study of the evolution of the relaxation time spectrum from liquid to solid state is in progress [70]. Preliminary results on vulcanizing polybutadienes indicate that the relaxation spectrum near the gel point is more complex than the simple spectrum presented in Eq. 3-6. In particular, the relation exponent n is not independent of the extent of reaction but decreases with increasing p.

4 Constitutive Modeling with the Critical Gel Equation

4.1 The Critical Gel Equation

Predictions using the observed relaxation time spectrum at the gel point are consistent with further experimental observations. Such predictions require a constitutive equation, which now is available. Insertion of the CW spectrum, Eq. 1-5, into the equation for the stress, Eq. 3-1, results in the linear viscoelastic constitutive equation of critical gels, called the 'critical gel equation'

$$\tau(t) = nS \int_{-\infty}^t (t - t')^{-(n+1)} \mathbf{C}^{-1}(t; t') \, dt' \quad \text{at } p = p_c. \tag{4-1}$$

It may alternatively be expressed with the rate of strain tensor $\partial \mathbf{C}_t^{-1}(t; t')/\partial t'$

$$\tau(t) = -S \int_{-\infty}^{t} (t - t')^{-n} \frac{\partial}{\partial t'} \mathbf{C}^{-1}(t; t') \, dt' \quad \text{at } p = p_c. \tag{4-2}$$

The cross-over to the glass at short times (or to other short-time behavior) is neglected here, which is justified as long as we only try to predict the long-time behavior, which is most affected by the solidification process.

The critical gel equation is expected to predict material functions in any small-strain viscoelastic experiment. The definition of 'small' varies from material to material. Venkataraman and Winter [71] explored the strain limit for crosslinking polydimethylsiloxanes and found an upper shear strain of about 2, beyond which the gel started to rupture. For percolating suspensions and physical gels which form a stiff skeleton structure, this strain limit would be orders of magnitude smaller.

4.2 Linear Viscoelastic Modeling of Critical Gels

With the gel equation, we can conveniently compute the consequences of the self-similar spectrum and later compare to experimental observations. The material behaves somehow in between a liquid and a solid. It does not qualify as solid since it cannot sustain a *constant* stress in the absence of motion. However, it is not acceptable as a liquid either, since it cannot reach a constant stress in shear flow at constant rate. We will examine the properties of the gel equation by modeling two selected shear flow examples. In shear flow, the Finger strain tensor reduces to a simple matrix with a shear component

$$(\mathbf{C}^{-1})_{12} = \gamma(t; t') \tag{4-3}$$

and a difference on the diagonal

$$(\mathbf{C}^{-1})_{11} - (\mathbf{C}^{-1})_{22} = (\gamma(t; t'))^2, \tag{4-4}$$

where

$$\gamma(t; t') = \int_{t'}^{t} \dot{\gamma}(t'') \, dt'' \tag{4-5}$$

is the shear strain between times t' and t. These components are inserted into Eq. 4-1 for calculating the shear stress and the first normal stress difference:

(a) *Startup of shear flow at constant rate.* An experiment is considered in which the material is initially kept at rest, $\dot{\gamma} = 0$, so that it can equilibrate completely. Starting at time $t = 0$, a constant shear rate $\dot{\gamma}_0$ is imposed. The resulting shear stress and normal stresses depend on the time of shearing. The shear stress response $\tau_{21}(t)$ of the critical gel is predicted as

$$\tau_{21}(t) = \dot{\gamma}_0 S \int_{0}^{t} (t - t')^{-n} \, dt' = \frac{1}{1 - n} \dot{\gamma}_0 S t^{1-n}. \tag{4-6}$$

The transient viscosity $\eta = \tau_{21}(t)/\dot{\gamma}_0$ diverges gradually without ever reaching steady shear flow conditions. This clarifies the type of singularity which the viscosity exhibits at the LST: The steady shear viscosity is undefined at LST, since the infinitely long relaxation time of the critical gel would require an infinitely long start-up time.

The corresponding first normal stress difference $N_1(t) = \tau_{11}(t) - \tau_{22}(t)$ as predicted from Eq. 4-2

$$N_1(t) = nS \int_{-\infty}^{t} (t - t')^{-(n+1)} \gamma(t; t')^2 \, dt' = \frac{2}{2-n} S\dot{\gamma}_0^2 t^{2-n} \qquad (4\text{-}7)$$

also grows with time without ever reaching a steady value. The ratio of first normal stress coefficient and viscosity

$$\frac{N_1}{\dot{\gamma}_0^2} \frac{\dot{\gamma}_0}{\tau_{21}} = \frac{2(1-n)}{2-n} t \qquad (4\text{-}8)$$

grows linearly with time. The relaxation exponent n solely determines the slope while the front factor cancels out. In experimental studies, the linear growth can be used as a convenient reference for finding the limits of linear response in this transient shear experiment (Fig. 13).

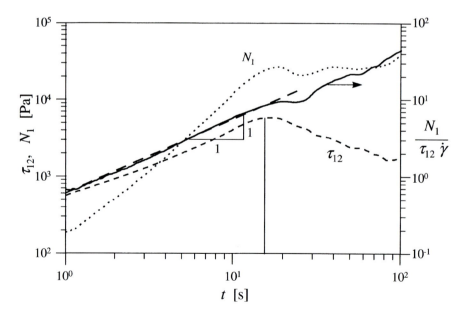

Fig. 13. Shear stress τ_{12} and first normal stress difference N_1 during start-up of shear flow at constant rate, $\dot{\gamma}_0 = 0.5 \text{ s}^{-1}$, for PDMS near the gel point [71]. The *broken line* with a slope of one is predicted by the gel equation for finite strain. The critical strain for network rupture is reached at the point at which the shear stress attains its maximum value

(b) *Oscillatory shear.* In a frequently used experiment, the sample is subjected to oscillatory shear at small amplitude γ_0. Prescribing a sinusoidal shear strain $\gamma(t)$ with an angular frequency, $\omega = 2\pi f$ [rad/s], which is defined by the number of cycles per time given by the frequency f [Hz],

$$\gamma(t) = \gamma_0 \sin(\omega t) \tag{4-9}$$

in Eq. 4-2 and determining the resulting shear stress

$$\tau_{21}(t) = G'(\omega)\gamma_0 \sin(\omega t) + G''(\omega)\gamma_0 \cos(\omega t) \tag{4-10}$$

results in the following functional form of the dynamic moduli [10], the storage modulus G' and the loss modulus G'', at the gel point

$$G'_c(\omega) = \frac{G''_c(\omega)}{\tan(n\pi/2)} = S\Gamma(1 - n)\cos(n\pi/2)\omega^n$$

$$\text{for } 0 < \omega < 1/\lambda_0, \, p = p_c. \tag{4-11}$$

Since G' and G'' are proportional to each other, the famous Cole–Cole plots [72] in which $\eta''(\omega)$ is plotted vs. $\eta'(\omega)$ [or $G'(\omega)$ is plotted vs. $G''(\omega)$] reduce to straight lines at the gel point.

The ratio of the two moduli is independent of frequency (Fig. 14)

$$\tan \delta_c = \frac{G''_c}{G'_c} = \tan \frac{n\pi}{2} \quad \text{for } 0 < \omega < 1/\lambda_0, \, p = p_c. \tag{4-12}$$

which means that the 'flat' phase angle is a direct measure of the relaxation exponent [7]:

$$n = \frac{2\delta_c}{\pi} \quad \text{for } 0 < \omega < 1/\lambda_0, \, p = p_c. \tag{4-13}$$

Related material functions are the complex modulus

$$G^*(\omega) = S\Gamma(1 - n)\omega^n \tag{4-14}$$

and the storage and loss compliance, $J'(\omega)$ and $J''(\omega)$

$$J'(\omega) = \frac{G'}{G'^2 + G''^2} = \frac{\cos n\pi/2}{S\Gamma(1 - n)}\omega^{-n} \tag{4-15}$$

$$J''(\omega) = \frac{G''}{G'^2 + G''^2} = \frac{\sin n\pi/2}{S\Gamma(1 - n)}\omega^{-n} \tag{4-16}$$

(c) *Creep and recovery behavior.* Similar is the modeling procedure for the strain in a creep experiment. The most simple creep recovery experiment

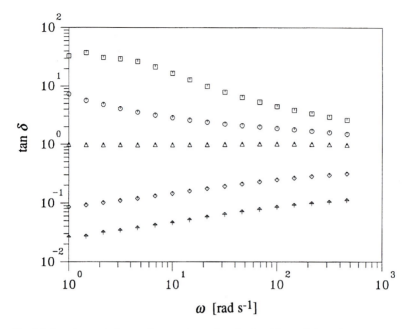

Fig. 14. Loss tangent of several stopped samples of vulcanizing polybutadiene ($M_w = 18\,000$) [31]. At the gel point, tan δ is frequency independent (flat curve in the middle). The relaxation exponent n can be easily evaluated from the data (tan $\delta = 1$ yields $n = 0.5$)

prescribes a pulse function

$$\tau_{21} = \begin{cases} 0 & \text{for } t < 0 \\ \tau_0 & \text{for } 0 < t < t_1, \\ 0 & \text{for } t_1 < t \end{cases} \tag{4-17}$$

where τ_0 is the applied shear stress and t_1 is the creep time. The strain response of any linear viscoelastic material

$$\gamma(t) = \int_{-\infty}^{t} J(t - t')\,\dot{\tau}(t')\,dt' \tag{4-18}$$

depends on the materials creep compliance $J(t - t')$. The classical relation between the creep compliance and the relaxation modulus [10]

$$t = \int_{0}^{t} G(t - s)\,J(s)\,ds \tag{4-19}$$

defines the creep compliance of the critical gel

$$J(t) = \frac{1}{S} \frac{\sin(n\pi)}{n\pi}\, t^n \quad \text{for } \lambda_0 < t. \tag{4-20}$$

The shear strain response has an analytical solution

$$\gamma(t) = \frac{\tau_0}{S} \frac{\sin(n\pi)}{n\pi} (t^n - (t - t_1)^n h(1 - t/t_1)) \quad \text{for } t > 0. \tag{4-21}$$

$h(x)$ is the Heaviside step function. It can be seen that neither the creep strain nor the strain rate will ever reach a steady value in finite times. When removing the stress, a complete recovery ($\gamma_\infty = 0$) is predicted for infinite times. This again is an example for the intermediate behavior between that of a liquid and a solid (Fig. 15).

(d) *Retardation Time Spectrum.* The relaxation behavior of critical gels can be represented equally well by the retardation time spectrum $L(\lambda)$ [73]. Both are related by

$$L(\lambda)H(\lambda) = \left(\left(\frac{1}{H(\lambda)} \int_0^\infty \frac{H(u)}{\lambda/u - 1} \frac{du}{u} \right)^2 + \pi^2 \right)^{-1} \tag{4-22}$$

We determine the long time end of the retardation spectrum by approximating Eq. 1-5 with

$$H(\lambda) = \frac{S}{\Gamma(n)} \lambda^{-n} \quad \text{for } 0 < \lambda < \infty \tag{4-23}$$

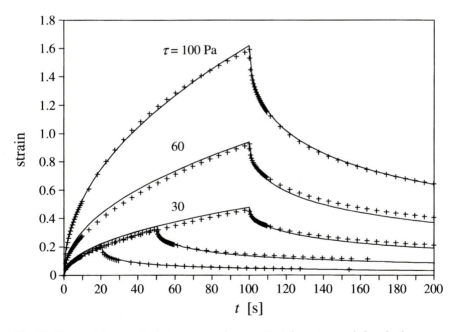

Fig. 15. Measured shear strain during creep under a constant shear stress and viscoelastic recovery after cessation of shear for **PDMS** near the gel point [71] plotted against the time. The *solid lines* are predicted by the gel equation for finite strain

due to lack of better information in the time range $0 < \lambda < \lambda_0$. Inserting Eq. 4-23 into Gross's relation, Eq. 4-22, leads to

$$L(\lambda) = \frac{\Gamma(n)}{S} \frac{\sin^2(n\pi)}{\pi^2} \lambda^n \quad \text{for } \lambda_0 < \lambda < \infty \tag{4-24}$$

We assume that the above solution is valid in about the same time range as the self-similar relaxation time spectrum, Eq. 1-5. The retardation time spectrum is also self-similar. It is characterized by its positive exponent n which takes on the same value as in the relaxation time spectrum.

5 Physical Gelation

The long range connectivity in the solidifying material may arise from physical phenomena instead of from chemical bonds. This process has been termed physical gelation. The large scale connectivity is meant in the sense that the motion of one molecule requires the motion of another molecule at considerable distance. This distance, called correlation length, increases with the advancement of the physical gelation process. The state at which the correlation length diverges defines the physical gel point. It is more difficult to define this gel point since in comparison with chemically crosslinking systems two of the most distinct criteria are missing: the molecular weight does not diverge and the system may be dissolved even after having passed the gel point.

At the beginning of the gelation process, more bonds are formed than are broken or dissolved. As a consequence, connectivity and correlation length grow in the material. However, this process cannot continue for long since the cluster size has a natural upper limit. This can be visualized by a simple argument. Let us consider a molecular cluster of N physical bonds with characteristic lifetime, λ_b. The average lifetime of this cluster is then λ_b/N and the survival probability is $\exp(-tN/\lambda_b)$. As the cluster grows (increasing N), its survival probability decreases. The limit is reached at a maximum average cluster size, N_{max}, at which the rate of bond breakage reaches the rate of bond formation. The characteristic time constant is then

$$\lambda_{pg} = \frac{\lambda_b}{N_{max}} \tag{5-1}$$

where the subscript pg stands for physical gel. The material has a corresponding longest relaxation time, λ_{max}. Early stages of cluster growth are governed by relaxation processes with a longest relaxation time which grows with the increasing connectivity. If λ_{max} exceeds λ_{pg}, then the cluster does not survive the relaxation process; stress is released by breakage of clusters. This type of relaxation process has been studied by Cates [74, 75].

We define a physical gel as a material which shows the gelation transition and has a λ_{max} after gelation which is orders of magnitude larger than before gelation. The characteristic equations at the gel point, Eqs. 1-1 and 1-5, need to be rewritten for a range of applicability $\lambda_0 < \lambda < \lambda_{pg}$. The critical gel equation, Eq. 4-1, also needs to be rewritten to accommodate this upper limit. The most simple way to do this is by inserting Eq. 1-5 with modified upper limit (∞ changed to λ_{pg}) into Eq. 3-3:

$$\tau(t) = nS \int_{-\infty}^{t} \left\{ \frac{\Gamma(n + 1, (t - t')/\lambda_{pg})}{\Gamma(n + 1)} \right\} (t - t')^{-(n+1)} \mathbf{C}^{-1}(t; t') dt' \qquad (5\text{-}2)$$

where $\Gamma(n + 1, (t - t')/\lambda_{pg})$ is an incomplete gamma function which is defined as

$$\Gamma(n + 1, x) = \int_{x}^{\infty} e^z z^n \, dz \qquad (5\text{-}3)$$

Strictly speaking, the physical gel at the gel point is still a liquid when observed at experimental times t_p which exceeds λ_{pg}. We therefore define a new dimensionless group, the gel number N_g

$$N_g = \frac{\lambda_{pg}}{t_p} = \frac{\text{lifetime of physical cluster}}{\text{process time}} \qquad (5\text{-}4)$$

The gelation transition is observable for $N_g > 10$. Otherwise, the material behaves as a liquid ($N_g < 1$). Little is known about materials near $N_g = 1$. For the following, we consider only materials with $N_g \gg 1$ and treat them just like chemical gels. The expression $\Gamma(n + 1, (t - t')/\lambda_{pg})/\Gamma(n + 1)$ in Eq. 5-2 approaches a value of one in this case of $N_g \gg 1$, and the critical gel equation, Eq. 4-1, is recovered. However, much work is needed to understand the role of non-permanent physical clusters on network formation and rheological properties.

The closest relation to chemical gelation is found with physical network systems in which the network junctions originate from some physical mechanism such as crystallization, phase separation, ionic bonds, or specific geometric complexes. Such systems have been reviewed recently by the Nijenhuis [76] and Keller [77]. Physical networks have the potential advantage that the junctions open or close when altering the environment (temperature, pressure, pH), i.e. the gelation process is reversible. The junctions, however, are less well-defined since their size and functionality (number of network strands which form a junction) varies throughout a sample. Their finite lifetime makes physical gels fluid-like in long-time applications, but it also allows them to heal if they get broken. The reversibility of junctions (and therefore connectivity) upon change of the environment in a physical gel is a characteristic feature which distinguishes it, for instance, from a highly entangled polymer melt or solution.

Beyond the notion of physical networks in which flexible strands are connected by junctions, we will use the term 'physical gelation' in the widest possible sense for polymeric systems which undergo liquid-solid transition due to any type of physical mechanism which is able to connect the polymer into large scale structures. These are quite manifold:

(a) Polymers lose their chain flexibility near their glass transition temperature and molecular motion correlates over longer and longer distances.

(b) Liquid crystalline polymers at their transition from nematic to smectic state gradually lose their molecular mobility.

(c) Suspensions in which the filler particles aggregate into sample spanning complexes.

The liquid-solid transition for these systems seems to have the same features as for chemical gelation, namely divergence of the longest relaxation time and power law spectrum with negative exponent.

Physical gelation responds strongly to stress or strain. The rate of bond formation, and therefore the growth rate of the correlation length, increases for some systems (increased rate of crystallization in semi-crystalline polymers; stress induced phase separation in block copolymers), but it also might decrease if the survival time of physical bonds is reduced by stress. Beyond the gel point, i.e. if the material is able to form sample spanning clusters which are characterized by multiple connectivity, physical gels are prone to creep under stress since bonds dissociate at a material-characteristic rate. This allows local relaxation of stress and reformation of physical bonds at reduced stress. In this case, breakage of a bond does not necessarily result in a reduction of the size of a cluster. Also, if stress is applied, the average lifetime of a bond, λ_b, decreases because of the energy input. This results in a reduction of λ_{pg}, which causes a reduction of the associated longest relaxation time, λ_{max}.

5.1 Physical Network Systems

Physical network systems, especially crystallizing polymers, represent the most widely investigated physically crosslinking macromolecular systems [68, 69, 78–81]. These systems comprise polymer melts and solutions in which network junctions are formed by small crystalline regions during the liquid-solid transition after cooling below the relevant melting temperature. The kinetics of this crystallization process depend on the degree of supercooling, ΔT, and rheological properties are also influenced by ΔT. A typical temperature profile for a crystallization experiment and the resulting evolution of fraction of crystalline polymer and dynamic moduli with time are shown schematically in Fig. 16. Although most crystallizing polymers exhibit the self-similar relaxation behavior at an intermediate state (Fig. 17 shows $\tan \delta$ at different times after cooling below T_m for a crystallizing polypropylene [68]), as commonly found in chemically crosslinking systems at the gel point, some systems showed no such power law relaxation at certain degrees of supercooling. This was attributed to non-uniform crystallization, which does not result in self-similar relaxation [69]. Also, the presence of melt state phase separation and residual high melting crystals can mask this characteristic relaxation pattern [68]. One of the

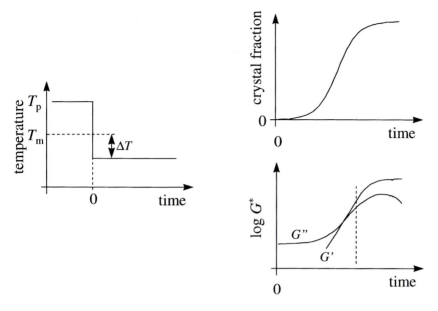

Fig. 16. Schematic of a typical temperature profile in a crystallization experiment and the resulting evolution of the fraction of crystalline polymer and dynamic moduli with time. The preheating temperature T_p is above the melting temperature T_m

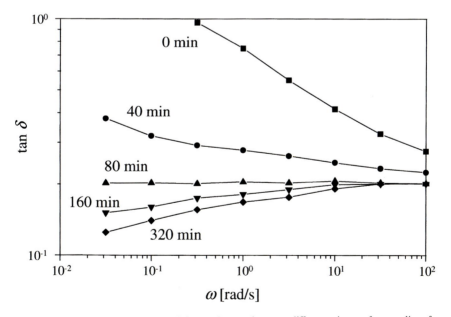

Fig. 17. Loss tangent of a crystallizing polypropylene at different times after cooling from $T = 100\,°C$ to $T = 40\,°C$ (below T_{melt}). Data from Lin et al. [68].

investigated systems, a bacterial polyester, followed the time-cure superposition principle in the liquid state up to the gel point [69]. This principle, however, could not be verified for samples beyond the gel point.

In general, low relaxation exponents n (between 0.1 and 0.25) are characteristic for crystallizing polymers, although some systems are known which show a stronger dependence on frequency ($n \approx 0.7 - 0.8$). A recent study on crystallizing isotactic polypropylene (iPP) resulted in a surprisingly low degree of crystallinity at the LST [80], which is a challenge to classical crystallization models. iPP forms a critical gel which is extremely soft (small S, large n).

Possible morphologies of partially crystalline polymers are shown in Fig. 18. Figure 18a depicts the case of small crystallites that act as physical crosslinks between polymeric chains, thus connecting those chains into a 3-dimensional network. In the case depicted in Fig. 18b, the material forms ribbon-shaped or needle-shaped crystalline regions in which different segments of a large number of chains are incorporated. This could explain the low degree of crystallinity at the LST as detected for the iPP system [80].

Reversible gelation is often encountered in bio-polymeric systems. Typical examples are solutions of polypeptide residues derived from animal collagen [82–84]. In these systems, ordered collagen-like triple helices form the physical crosslinks.

Microphase separated systems are also known to yield a physical network which results in the self-similar relaxation pattern at an intermediate state

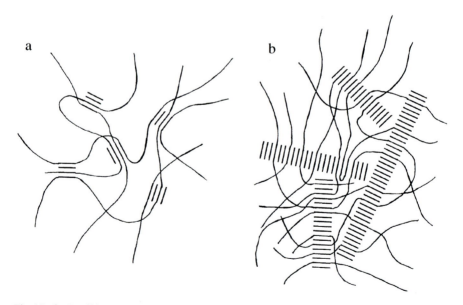

Fig. 18a, b. Possible morphologies of partially crystalline polymers. Small crystallites act as crosslinks (**a**); large ribbon-shaped or needle-shaped crystalline regions connect a large number of polymeric chains (**b**)

corresponding to the LST. Examples in which such behavior was observed are a segmented polyurethane elastomer with liquid crystalline hard segments (in which the phase-separated submicron mesophase acts as a provider of crosslinks below the isotropization temperature of the mesophase [85]) and several diblock and triblock copolymers with microphase separation below the ODT temperature [86, 87]. A styrene-isoprene-styrene triblock copolymer system, for example, showed rapid microphase separation after cooling below the ODT temperature, while the large scale spatial order, resulting in a physical network, needed long annealing times. It was recently suggested that the disappearance of a terminal zone behavior, as encountered in block copolymers below the ODT temperature, is generally true for any polymer with layered structure [87].

Recently, it was shown that polydimethylcarbosiloxanes with a small content of side carbonyl groups (PDMS-C) exhibit increasing viscosity and formation of a physical network at elevated temperatures [88, 89]. This was attributed to a rearrangement of intramolecular hydrogen bonds, which formed between the carboxyls during the synthesis and isolation of the polymers, forming intermolecular hydrogen bonds.

5.2 Dynamic Glass Transition

A glass transition is introduced dynamically when probing polymer molecules on such short time scales (in a high frequency dynamic experiment, for instance) that conformational rearrangements have no time to develop. The spectrum of polymers in this dynamic glass transition region is given by a power law with negative exponent. This has already been realized by Tobolsky [90], who introduced a wedge-box spectrum to describe entangled polymer melts where the 'wedge' (power law with negative slope) represented the relaxation behavior in the glass transition region. At long times this behavior is masked by the onset of the entanglement and/or flow regime. The glass transition spectrum looks like the CW spectrum for the critical gel except that it is cut off at a characteristic relaxation time, λ_{char}, i.e. the longest relaxation time is finite. This suggests that a polymer at the glass transition might be considered in the physical gel framework.

The amorphous solid state may be viewed as an extension of the liquid phase below a characteristic temperature. For low molecular weight materials, the atoms are frozen into their relative position at low temperatures [91]; this is called configurational freezing. For high molecular weight materials such as polymers, the molecular conformations freeze in and arrest molecular motion (conformational freezing); the temperature of conformational freezing is called glass transition temperature. T_g. T_g is above the temperature of configurational freezing. The conformational freezing results in an increase of the correlation length for molecular motion. The divergence of the correlation length denotes the instant of solidification; thus the relaxation of the material is similar to that of a physical gel. The cut-off time $\lambda_{char}(T)$ of the relaxation spectrum depends on temperature and becomes very long when the temperature is lowered towards the glass transition temperature, T_g, but it remains finite at T_g. Dynamic mechanical experiments near T_g are needed for exploring the applicability of the gelation framework here. A very comprehensive study of dynamic mechanical behavior near T_g has been given by Zorn et al. [92].

The glass transition involves additional phenomena which strongly affect the rheology: (1) Short-time and long-time relaxation modes were found to shift with different temperature shift factors [93]. (2) The thermally introduced glass transition leads to a non-equilibrium state of the polymer [10]. Because of these, the gelation framework might be too simple to describe the transition behavior.

5.3 Liquid Crystalline Polymers at their Nematic-Smectic Transition

Viscoelastic response of liquid crystalline polymers (LCP) is very sensitive to smectic-nematic and smectic-isotropic phase transitions. Typical side chain LCPs with mesogenic groups pendant to flexible backbones show liquid-like relaxation behavior at low frequencies in the nematic state, i.e. the storage modulus, G', is proportional to frequency, ω, and the loss modulus, G'', is proportional to ω^2. At intermediate frequencies, a power law dependence best describes the dynamic moduli [94, 95]. Chemically crosslinking polymers below the gel point show the same behavior, which is followed by a transition to entanglement and/or glass transition regime at higher frequencies. LCPs in the smectic phase do not exhibit a low frequency drop-off to liquid-like behavior, at least not in the experimentally observable frequency regime. G' and G'' seem to level off at low frequency, suggesting a more solid-type relaxation behavior. No real power-law dependence is observed in the smectic mesophase; however, at intermediate frequencies indications of self-similar relaxation can be observed.

5.4 Suspensions

Transition from liquid behavior to solid behavior has been reported with fine particle suspensions with increased filler content in both Newtonian and non-Newtonian liquids. Industrially important classes are rubber-modified polymer melts (small rubber particles embedded in a polymer melt), e.g. ABS (acrylo-nitrile-butadiene-styrene) or HIPS (high-impact polystyrene) and fiber-reinforced polymers. Another interesting suspension is present in plasticized polyvinylchloride (PVC) at low temperatures, when suspended PVC particles are formed in the melt [96]. The transition becomes evident in the following experimental observations:

- The limiting storage modulus (at low frequencies) and relaxation modulus (at long times) become finite at high concentration, while they are zero at low concentration [97–102].
- The zero-shear viscosity and the dynamic viscosity (at low frequencies) diverge at high concentration, while they are constant at low concentration [99, 100, 102–105].

Liquid-solid transitions in suspensions are especially complicated to study since they are accompanied by additional phenomena such as order-disorder transition of particulates [98, 106, 107], anisotropy [108], particle-particle interactions [109], Brownian motion, and sedimentation-particle convection [109]. Furthermore, the size, size distribution, and shape of the filler particles strongly influence the rheological properties [108, 110]. More comprehensive reviews on the rheology of suspensions and rubber modified polymer melts were presented by Metzner [111] and Masuda et al. [112], respectively.

Oscillatory shear experiments are the preferred method to study the rheological behavior due to particle interactions because they directly probe these interactions without the influence of the external flow field as encountered in steady shear experiments. However, phenomena that arise due to the external flow, such as shear thickening, can only be investigated in steady shear experiments. Additionally, the analysis is complicated by the different response of the material to shear and extensional flow. For example, very strong deviations from Trouton's ratio (extensional viscosity is three times the shear viscosity) were found for suspensions [113].

We expect the liquid/solid transition to express itself in the same general relaxation patterns as chemical gelation, with a self-similar relaxation time spectrum at the gel point. A starting point for this hypothesis is the work of Castellani and Lomellini [102, 114], who compared the rheology of rubber-modified thermoplastics with increasing rubber content to the behavior during physical gelation. The relaxation spectra of ABS with different PBD content presented by Masuda et al. [112] also suggest this approach, since they seem to contain a power law at long times.

6 Rheometry Near the Gel Point

Viscoelastic properties on intermediate time scales are most appropriate for studying gelation. The stretching of the spectrum in the approach of the gel point (from either side), and the self-similarity of the spectrum at the gel point can best be observed by forcing the material through the transition while simultaneously measuring its continuously evolving linear viscoelastic properties. Small strain tests are preferable, since they avoid rupturing the fragile network structure. No specific equipment is required beyond what one would use for characterizing viscoelasticity in liquids or solids. Most common are rotational rheometers with concentric disk fixtures, cone and plate fixtures, or Couette geometry. Samples have to be prepared in the rheometer fixtures since they are too fragile and too sticky to be transferred later.

6.1 Oscillatory Shear

Small amplitude oscillatory shear is the method of choice for materials with very broad distributions of relaxation modes, such as materials near LST, and for materials which undergo change during the measurement. The dynamic moduli in Eq. 4–10 are defined by [10]

$$G'(\omega) = G_e + \omega \int_0^\infty [G(t) - G_e] \sin(\omega t)\, dt = \text{storage modulus}, \qquad (6\text{-}1)$$

$$G''(\omega) = \omega \int_0^\infty G(t) \cos(\omega t)\, dt \qquad\qquad = \text{loss modulus}. \qquad (6\text{-}2)$$

The above equations are generally valid for any isotropic material, including critical gels, as long as the strain amplitude γ_0 is sufficiently small. The material is completely characterized by the relaxation function $G(t)$ and, in case of a solid, an additional equilibrium modulus G_e.

The basic advantages of small amplitude oscillatory strain (shear or extension) come through its spectroscopic character, the experimental time for taking a single data point being roughly equal to the period of the strain wave, $2\pi/\omega$. This allows to measure specific relaxation modes with time constants in the order of $1/\omega$ independently of any longer or shorter modes which might be present in the polymer, i.e. only a small fraction of the spectrum is actually sampled. This is shown in Fig. 19, where a single power law spectrum with negative exponent ($n = 0.7$) was used to calculate the integral kernels of the following equations at $\omega = \lambda/\lambda_0$.

$$G'(\omega) = \int_0^{\lambda_{max}} H(\lambda) \frac{(\omega\lambda)^2}{1 + (\omega\lambda)^2} \frac{d\lambda}{\lambda} \qquad (6\text{-}3)$$

$$G''(\omega) = \int_0^{\lambda_{max}} H(\lambda) \frac{\omega\lambda}{1 + (\omega\lambda)^2} \frac{d\lambda}{\lambda} \qquad (6\text{-}4)$$

The divergence of the longest relaxation time does not perturb the measurement. In comparison, steady state properties (the steady shear viscosity, for instance) would probe an integral over all relaxation modes and, hence, fail near the gel point.

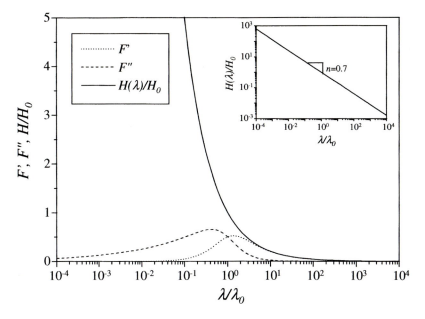

Fig. 19. $F'(\lambda/\lambda_0)$ and $F''(\lambda/\lambda_0)$ for the evaluation of G' and G'' at $\omega = 1/\lambda_0$.

$$G'(\omega = 1/\lambda_0) = H_0 \int_0^\infty F'(x)dx \quad \text{with } F'(x) = x^{-(n+1)}\frac{x^2}{1+x^2}$$

$$G''(\omega = 1/\lambda_0) = H_0 \int_0^\infty F''(x)dx \quad \text{with } F''(x) = x^{-(n+1)}\frac{x}{1+x^2}$$

Insert: Reduced power law spectrum $H(\lambda) = H_0(\lambda/\lambda_0)^{-n}$

Typical for the spectroscopic character of the measurement is the rapid development of a quasi-steady state stress. In the actual experiment, the sample is at rest (equilibrated) until, at $t = 0$, oscillatory shear flow is started. The shear stress response may be calculated with the general equation of linear viscoelasticity [10] (introducing Eqs. 4-3 and 4-9 into Eq. 3-2)

$$\tau_{21}(t) = \int_{-\infty}^0 0 + \int_0^t G(t - t')\omega\gamma_0 \cos(\omega t')dt' \tag{6-5}$$

The first integral denotes the rest period, $-\infty < t' < 0$, where the strain rate is zero. The second integral contains a relaxation function which we chose very broad, including relaxation times much larger than the period $2\pi/\omega$. Integration and quantitative analysis clearly showed (without presenting the detailed figures here) that the effect of the start-up from rest is already very small after one cycle and definitely is negligible after two cycles. Eq. 6-5 simplifies to

$$\tau_{21}(t) \approx \int_{-\infty}^{t} G(t - t')\omega\gamma_0 \cos(\omega t')\,dt' \tag{6-6}$$

The start-up time does not depend on the longest relaxation time of the material even if it is orders of magnitude larger than the period $2\pi/\omega$ [115]. This is an important prerequisite for an experiment near LST.

The dynamic mechanical experiment has another advantage which was recognized a long time ago [10]: each of the moduli G' and G'' independently contains all the information about the relaxation time distribution. However, the information is weighted differently in the two moduli. This helps in detecting systematic errors in dynamic mechanical data (by means of the Kramers-Kronig relation [54]) and allows an easy conversion from the frequency to the time domain [8, 116].

Figures 20 and 21 show typical dynamic moduli and loss tangent distributions. These were measured by Chambon and Winter [5] on several partially crosslinked samples of PDMS.

Limitations of the experiment at low frequencies come from the long experimental times, during which the sample structure may change so much that the entire experiment becomes meaningless. At high frequencies, limitations

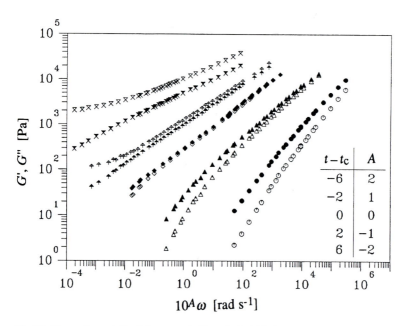

Fig. 20. Dynamic moduli, G' (*open symbols*) and G'' (*filled symbols*), for five partially crosslinked PDMS samples at different extents of reactions (same data as Fig. 2). G' curves downward for the liquid and curves to the left for the solid. The *straight lines* belong to the sample which is very close to the gel point. $t - t_c$ is given in minutes

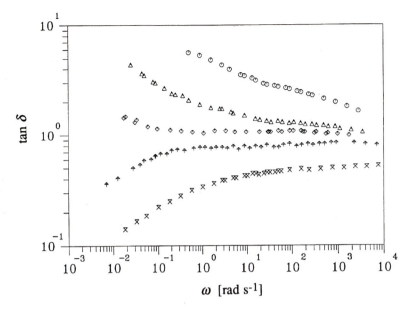

Fig. 21. Loss tangent, tan δ, for five partially crosslinked PDMS samples at different extents of reactions. For the liquid, a negative slope is observed. At the LST, tan δ is independent of frequency. The loss tangent of the solid material exhibits a positive slope

arise from inertial effects. The effect of changes in the sample will be discussed next.

6.2 Sample Mutation

One really would like to know G' and G'' data over an as wide as possible frequency window of the sample at intermediate states during the transition. This has been achieved best for chemical gelation by stopping the reaction at intermediate extents p with a catalyst poison [5] and then probing the stable samples (see Figs. 20 and 21). However, this stopping of the crosslinking is only possible for exceptional materials. More common is the situation where the transition process cannot be brought to a halt. The solidifying material has to be probed without stopping the reaction. Each data point in a sequence of measurements, having a sampling time Δt, belongs to a different state of the material. This is displayed schematically in Fig. 22. Since the properties at these intermediate frequencies evolve continuously, interpolation may be used to determine the properties at any time in between the measurements. The data are interpolated to obtain the material properties at discrete material states. By this procedure [58, 117, 118], a range of properties is available at discrete states of the evolving material structure.

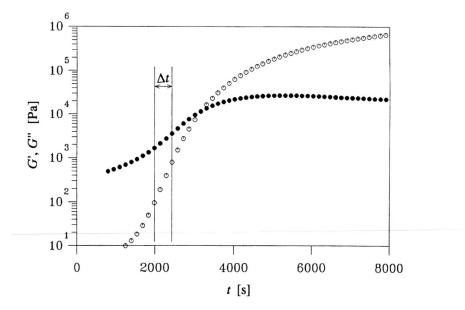

Fig. 22. Schematic evolution of dynamic moduli during crosslinking. The time period Δt corresponds to the time necessary for a rheological measurement in case of a reacting sample. If the reaction is stopped, Δt corresponds to the time in which the poison diffuses into the sample. The reaction is still carried on until the entire sample is poisoned. Then, the moduli remain constant

The effect of mutation is different in case of stopped samples, but the phenomenon cannot be completely avoided. Here, the experimental time period Δt is determined by the poison diffusion. The catalyst poison solution is sprayed on top of a reacting sample and then diffuses into the core of the sample where it stops the reaction sequentially layer by layer. This leads to small inhomogeneity in the sample, since the reaction near the upper surface is stopped earlier than the reaction near the bottom of the mold.

Sample changes during the measurement might cause severe problems. The shear stress response of a crosslinking system exemplifies this nicely:

$$\tau_{21}(t) = \int_{-\infty}^{t} G_{t''=t'}^{t''=t}[t, t', p(t'')] \omega \gamma_0 \cos(\omega t') dt' \tag{6-7}$$

The stress depends on the extent of reaction, $p(t')$, which progresses with time. However, it is not enough to enter the instantaneous value of $p(t')$. Needed is some integral over the crosslinking history. The solution of the mutation problem would require a constitutive model for the fading memory functional $G_{t''=t'}^{t''=t}[t, t', p(t'')]$, which is not yet available. This restricts the applicability of dynamic mechanical experiments to slowly crosslinking systems.

The mutation number [117, 119]

$$N_{mu} = \Delta t \frac{1}{g} \frac{\partial g}{\partial t} \qquad (6-8)$$

gives an estimate of the relative changes of a measured variable g (standing here for G' or G'') during the sampling time Δt. The spectroscopic character of dynamic mechanical experiments somehow alleviates the mutation problem. Mutation effects are considered negligible if the moduli G' and G'' change by 10% or less ($N_{mu} < 0.1$) during the experimental time, $2\pi/\omega$. Equivalent criteria apply to all the various material functions. The decisive dimensionless group (mutation number) has to stay below an acceptable tolerance level, $N_{mu} < 0.1$.

We mostly chose to probe each frequency individually to minimize the strain on the material and to expand the available frequency window. The experimental time can be reduced by simultaneously applying the sinusoidal strains of the lowest frequencies [120] and then quickly adding the higher frequency part of the spectral probing.

Time-resolved measurements on the changing sample have the advantages that the critical gel properties can be obtained from a single experiment and that a value for the rate of evolution of properties comes with the data.

6.3 Time-Temperature Superposition

Time-temperature superposition [10] increases the accessible frequency window of the linear viscoelastic experiments. It applies to stable material states where the extent of reaction is fixed ('stopped samples'). Winter and Chambon [6] and Izuka et al. [121] showed that the relaxation exponent n is independent of temperature and that the front factor (gel stiffness) shifts with temperature

$$S(T) = S(T_0) \frac{a_T{}^n}{b_T}. \qquad (6-9)$$

This behavior is in between that of a liquid and a solid. As an example, PDMS properties obey an Arrhenius-type temperature dependence because PDMS is far above its glass transition temperature (about $-125°C$). The temperature shift factors are

$$a_T = \exp\left\{ \frac{E}{R}\left(\frac{1}{T} - \frac{1}{T_0} \right) \right\}; \; b_T = \frac{\rho(T_0)T_0}{\rho(T)T}. \qquad (6-10)$$

From the PDMS shift factors determined by Winter and Chambon, one may estimate that room temperature fluctuations affect the gel strength by no more than 5%.

Time-temperature superposition at the gel point does not let us distinguish between the vertical and the horizontal shift, since the spectra are given by straight lines in a log/log plot. We can only determine the total product a_T^n/b_T [121] without being able to break it down into separate values a_T and b_T. For polymers at temperatures far above the glass transition temperature T_g, however, the crosslinking does not seem to affect the shift factors in any major way and a_T and b_T values may approximately be taken from the temperature dependence of the precursor viscosity and the modulus of the fully crosslinked system, respectively. This has been found for crosslinking polycaprolactones far above the glass transition temperature [121]. More experiments are needed to confirm this very simple relation. It is expected to loose its validity for samples for which crosslinking strongly increases T_g.

6.4 Time-Cure Superposition

Measurement of the equilibrium properties near the LST is difficult because long relaxation times make it impossible to reach equilibrium flow conditions without disruption of the network structure. The fact that some of those properties diverge (e.g. zero-shear viscosity or equilibrium compliance) or equal zero (equilibrium modulus) complicates their determination even more. More promising are time-cure superposition techniques [15] which determine the exponents from the entire relaxation spectrum and not only from the diverging longest mode.

Adolf and Martin [15] postulated, since the near critical gels are self-similar, that a change in the extent of cure results in a mere change in scale, but the functional form of the relaxation modulus remains the same. They accounted for this change in scale by redefinition of time and by a suitable redefinition of the equilibrium modulus. The data were rescaled as $G'/G_e(p)$ and $G''/G_e(p)$ over $\omega\lambda_{max}(p)$. The result is a set of master curves, one for the sol (Fig. 23a) and one for the gel (Fig. 23b).

Time-cure superposition is valid for materials which do not change their relaxation exponent during the transition. This might be satisfied for chemical gelation of small and intermediate size molecules. However, it does not apply to macromolecular systems as Mours and Winter [70] showed on vulcanizing polybutadienes.

6.5 Growth Rate of Moduli

The rate of change through the transition has not been studied widely. However, the growth of G' and G'' due to the increasing network connectivity seems to follow a regular pattern. For all our experiments of that type which were restricted by the accessible frequency range of the rheometer, the growth rate of G' at the gel point was typically twice as high as that of G'':

$$\left(\frac{1}{G'}\frac{\partial G'}{\partial p}\right)_\omega \cong 2\left(\frac{1}{G''}\frac{\partial G''}{\partial p}\right)_\omega \quad \text{for finite } \omega. \tag{6-11}$$

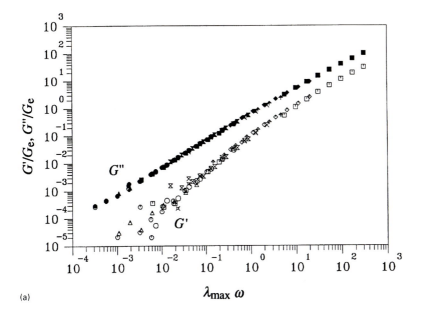

(a)

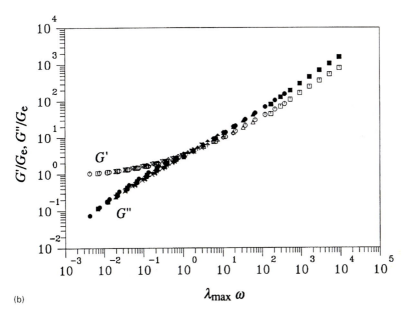

(b)

Fig. 23. Master curve obtained by time cure superposition of data on curing epoxy (a) before the LST and (b) after the LST [15]

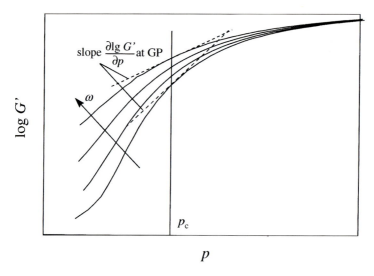

Fig. 24. Schematic of the evolution of the dynamic storage modulus, G', with extent of reaction, p, at four different frequencies

The growth rate decreases with frequency. Figure 24 shows this schematically for the storage modulus. The frequency dependence is the same for both G' and G'', and it also follows a power law (Fig. 25).

$$\left(\frac{1}{G'}\frac{\partial G'}{\partial p}\right)_\omega \propto \left(\frac{1}{G''}\frac{\partial G''}{\partial p}\right)_\omega \propto \omega^{-\kappa} \quad \text{for } p \text{ near } p_c. \tag{6-12}$$

Precise knowledge of the critical point is not required to determine κ by this method because the scaling relation holds over a finite range of p at intermediate frequency. The exponent κ has been evaluated for each of the experiments of Scanlan and Winter [122]. Within the limits of experimental error, the experiments indicate that κ takes on a universal value. The average value from 30 experiments on the PDMS system with various stoichiometry, chain length, and concentration is $\kappa = 0.214 \pm 0.017$. Exponent κ has a value of about 0.2 for all the systems which we have studied so far. Colby et al. [38] reported a value of 0.24 for their polyester system. It seems to be insensitive to molecular detail. We expect the dynamic critical exponent κ to be related to the other critical exponents. The frequency range of the above observations has to be explored further.

6.6 Inhomogeneities

The crosslinking process should not be considered to be completely homogeneous. Several phenomena might cause inhomogeneities. On a molecular scale, we

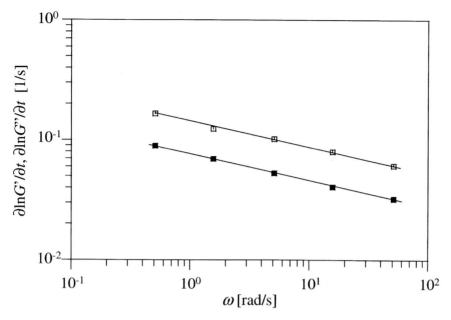

Fig. 25. Frequency dependence of moduli growth rate at the LST of crosslinking PDMS. Data from Scanlan and Winter [58]

can expect that fluctuations in crosslink density grow by excluding mobile sol molecules from regions of high crosslink density. This has been observed for highly crosslinked polymers and might occur at early stages of crosslinking.

Wall effects might also lead to inhomogeneities on the molecular level. Two possibilities can be envisioned: exclusion of large molecular clusters from the layer adjacent to a solid boundary, or adhesion of the largest clusters to the wall (while smaller molecules come and go near the wall).

These small-scale inhomogeneities would not be visible in rheological experiments since they average out in a macroscopic approach. However, there are inhomogeneities which do affect the macroscopic experiment. Most easily encountered is a temperature gradient which leads to gradients in the rate of reaction. The hotter region of the sample reaches the gel point first and the entire transition phenomenon gets smeared out. A clear gel point cannot be detected macroscopically. We have encountered this phenomenon quite often and use the purity of the self-similar behavior as an important criterion for a well-defined experiment.

Another typical example of inhomogeneity in rheometry is the oxidation of a polymer in a rotational rheometer in which a disk-shaped sample is held between metal fixtures. The oxygen enters the sample through the free surface (at the outer diameter) and diffuses radially inwards. The result is a radial gradient in properties which changes with time. If the reaction with oxygen results in a solidification, then the outer edge of the sample solidifies much before the inner regions, thus leading to an experiment-inherent inhomogeneity of the sample.

7 Detection of the Liquid-Solid Transition

The gel point is reached when the largest molecular or supermolecular cluster diverges to infinity [123]. This cannot be measured directly, but the consequences are so dramatic that they can be seen in a wide range of phenomena. Detection methods involve static light scattering on diverging clusters and dynamic light scattering with diverging correlation time [124–128], dissolving in a good solvent and extraction of a sol fraction [129], permeability [130], nuclear magnetic resonance spectroscopy [131], differential scanning calorimetry [132–134], and infrared [29, 135] and Raman spectroscopy [136, 137]. Very sensitive indicators are rheological properties. Viscosity, first normal stress difference coefficient, equilibrium compliance, and longest relaxation time diverge at the gel point and the distribution of relaxation modes adopts a power law. These rheological features have been used extensively for detecting the gel point, and the following discussion will focus on rheological methods.

During our early experiments on chemical gels, when first observing the intermediate state with the self-similar spectrum, Eq. 1–5, we simply called it 'viscoelastic transition'. Then, numerous solvent extraction and swelling experiments on crosslinking samples showed that the 'viscoelastic transition' marks the transition from a completely soluble state to an insoluble state. The sol-gel transition and the 'viscoelastic transition' were found to be indistinguishable within the detection limit of our experiments. The most simple explanation for this observation was that both phenomena coincide, and that Eqs. 1-1 and 1-5 are indeed expressions of the LST. Modeling calculations of Winter and Chambon [6] also showed that Eq. 1-1 predicts an infinite viscosity (see Sect. 4) and a zero equilibrium modulus. This is consistent with what one would expect for a material at the gel point.

Physical solidification processes have no criterion for the gel point which would be as decisive as the molecular weight divergence of the sol-gel transition. LST measurements involve elaborate observations of flow to non-flow transition, and they rely on subjective judgement. The broadening of the spectrum into a power law distribution seems to coincide with such cessation of flow observations. We have therefore suggested that it should be valid to generalize from the behavior of chemical gels and identify LSTs of physical gels also by the CW spectrum [67]. In this way we can clearly identify both the approach to gelation and the transition of materials which otherwise would lack a clear definition.

7.1 Diverging Rheological Properties

The diverging rheological properties are an unambiguous sign of the approaching gel point even if the measurement breaks down in its immediate vicinity (see Fig. 5). Macosko [138] gave a well-balanced overview of the methods up to 1985. The most common rheological tests for detecting LST measure the divergence of the steady shear viscosity [139–147] or the appearance of an equilibrium modulus [143, 146–148]. The equilibrium modulus evades accurate measurement near LST since its value is zero at LST and remains below the detection limit for a considerable time. It appears in a stress relaxation experiment as the long time limit of the relaxation function or in oscillatory strain as the low frequency asymptote of G'.

Measurement of the diverging steady shear viscosity is an appealing experiment because of its simplicity. Even the torque on a processing machine might serve as an estimate of the diverging viscosity. It has, however, severe disadvantages that need to be considered [149]:

1. Near LST, the relaxation times become very long, and steady shear flow cannot be reached in the relatively short transient experiment. Large strains are the consequence for most reported data.
2. At large strain, the liquid shows shear thinning in some poorly understood fashion. Shearing causes breakage of the fragile network structure near LST, which has been observed as an apparent delay in gelation.
3. LST is found by extrapolation. The actual experiment may also end prematurely some time before LST if the developing structure in the material is very stiff and the rising stress overloads the rheometer.

The diverging viscosity, therefore, does not show the real gel point. The transition may appear early because of torque overload or it may be delayed by chain scission due to large strain. This apparent gel point, however, is still important since it relates to processing applications in which either the machine would clog or the newly formed network structure would break (or both).

7.2 Monotonously Changing Properties

The relaxation modulus evolves gradually during gelation. A set of data along the lines of Fig. 2 gives a good estimate of where the gel point occurs. The problem with it is that one cannot decide very well when exactly $G(t)$ has straightened out into a power law.

Dynamic mechanical properties also evolve gradually during the LST of polymeric systems. The gel point is reached when tan δ becomes independent of frequency [58, 63, 65, 120, 149, 150] (see Eq. 4-12). Lines of tan $\delta(t)$ at several frequencies $\omega_1, \omega_2, \omega_3, \omega_4, \omega_5$, etc. decay gradually and intersect at the gel point (see Fig. 26). The method is very effective. The instant of gelation can be measured as precisely as the accuracy of the rheometer permits – a significant advantage over extrapolation methods. An additional advantage is that the strain is kept small and shear modification of the molecular structure is avoided. The experiment not only tags the instant of gelation but also provides the value of the relaxation exponent n (see Eq. 4-12).

Even before reaching the gel point, the converging lines can be extrapolated towards the expected gel point. The LST can be anticipated. This is convenient for preparing materials somewhere before the gel point but very close to it. The cross-linking can be stopped at a defined distance before LST, $(p_c - p)$.

There are also some far-fetched proposals for the LST: a maximum in tan δ [151] or a maximum in G'' [152] at LST. However, these expectations are not consistent with the observed behavior. The G'' maximum seems to occur much beyond the gel point. It also has been proposed that the gel point may be reached when the storage modulus equals the loss modulus, $G' = G''$ [153, 154], but this is contradicted by the observation that the $G' - G''$ crossover depends on the specific choice of frequency [154]. Obviously, the gel point cannot depend on the probing frequency. Chambon and Winter [5, 6], however, showed that there is one exception: for the special group of materials with a relaxation exponent value $n = 0.5$, the loss tangent becomes unity, tan $\delta_c = 1$, and the $G' - G''$ crossover coincides with the gel point. This shows that the crossover $G' = G''$ does *not* in general coincide with the LST.

7.3 Uniqueness of tan δ Method

A self-similar relaxation spectrum with a negative exponent (-n) has the property that tan δ is independent of frequency. This is convenient for detecting the instant of gelation. However, it is not evident that the claim can be reversed. There might be other functions which result in a constant tan δ. This will be analyzed in the following.

A constant loss tangent, tan $\delta \neq f(\omega)$, requires dynamic moduli

$$G' = h(\omega), \quad G'' = Ah(\omega) \rightarrow \tan \delta = G''/G' = A = \text{const.} \tag{7-1}$$

where $h(\omega)$ may be an arbitrary function. The question is whether $h(\omega)$ has to be a power law $h(\omega) \propto \omega^n$ or whether other functions are permitted. If other functions are permitted, then tan $\delta \neq f(\omega)$ (for $1/\lambda_0 < \omega < \infty$) is not a unique criterion for the LST. For the following derivation, the short time behavior can be neglected since it has no effect on the ensuing argument. For simplicity we assume that the above condition applies over the entire frequency range, $0 < \omega < \infty$. The Kramers-Kronig relation

$$G'(\omega) = \frac{2}{\pi} \int_0^\infty \frac{G''(x)}{1 - (x/\omega)^2} \frac{dx}{x} \tag{7-2}$$

can be evaluated for the above function

$$h(\omega) = \frac{2}{\pi} \int_0^\infty \frac{Ah(x)}{1 - (x/\omega)^2} \frac{dx}{x} \tag{7-3}$$

Substitution of $y = x/\omega$ gives

$$1 = A\frac{2}{\pi} \int_0^\infty \frac{h(y\omega)}{h(\omega)} \frac{1}{1 - y^2} \frac{dy}{y} \tag{7-4}$$

Any function of $y\omega$ with the factorization property

$$h(y\omega) = k(y)h(\omega), \tag{7-5}$$

where $k(y)$ may be an arbitrary function, is solution since it reduces the integral to some constant which defines A:

$$1 = \frac{2A}{\pi} \int_0^\infty \frac{k(y)}{1 - y^2} \frac{dy}{y} \tag{7-6}$$

A power law $h(\omega) \propto \omega^n$ satisfies this condition, Eq. 7-5, but any number of other functions with that property might be invented. The $\tan \delta$ criterion, therefore, might be not unique. However, no other material has yet been found which also obeys $\tan \delta = $ constant in the terminal frequency region, and we suggest the continued use of the $\tan \delta$ method for detecting LST until a counter example can be found.

7.4 Determination of S and n

Dynamic mechanical data near the gel point allow easy determination of the parameters of the critical gel, Eq. 1-1. $\tan \delta$, as shown in Fig. 26, gives the relaxation exponent n

$$n = \frac{2}{\pi} \tan^{-1} \left(\frac{G''}{G'} \right) \tag{7-7}$$

$G''/G' = (G''/G')_c$ is the value where the curves intersect in a single point. The same data can be rearranged into

$$S = \frac{G'(\omega)}{\omega^n \cos(n\pi/2)\Gamma(1 - n)} \tag{7-8}$$

S has to be evaluated at the gel point (with $G' = G'_c(\omega)$ at low frequencies). This completely characterizes the critical gel. The critical gel behavior is valid above a material characteristic time constant λ_0. The relation between S and n given by Eq. 3-4 holds only at the LST.

The above two equations are generally valid for viscoelastic liquids and solids. In this case, n and S would depend on frequency. In this sense, the above

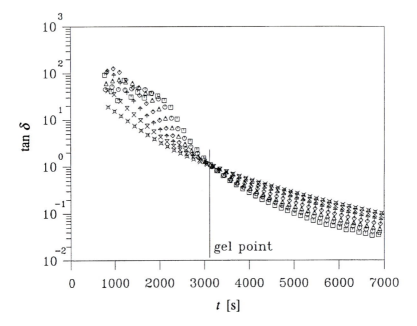

Fig. 26. Tan δ of a crosslinking PBD ($M_w = 18\,000$) as a function of reaction time [31]. Parameter is the frequency ω. The polymer is vulcanized at the pendant vinyl units with a bifunctional silane crosslinker using a platinum compound as catalyst. The curves intersect at the gel point resulting in tan $\delta \neq f(\omega)$

definitions are also valid in the vicinity of the gel point. However, n and S are only independent of frequency at the gel point where self-similar relaxation governs the rheological behavior.

8 Other Observations of Power Law Relaxation

Power law relaxation is no guarantee for a gel point. It should be noted that, besides materials near LST, there exist materials which show the very simple power law relaxation behavior over quite extended time windows. Such behavior has been termed 'self-similar' or 'scale invariant' since it is the same at any time scale of observation (within the given time window). Self-similar relaxation has been associated with self-similar structures on the molecular and supermolecular level and, for suspensions and emulsions, on particulate level. Such self-similar relaxation is only found over a finite range of relaxation times, i.e. between a lower and an upper cut-off, λ_1 and λ_u. The exponent may adopt negative or positive values, however, with different consequences and limitations. A LST would require additional characteristics such as no upper time limit for the self-similar region, stretching out of the spectrum at the approach of the gel point and shrinking beyond the gel point, and different curvature of the storage modulus before and after the gel point. These characteristics have not been found in the following examples.

8.1 Self-Similar Relaxation with Negative Exponent Value

For negative exponent values, the symbol $-n$ with $n > 0$ will be used. The self-similar spectrum has the form

$$H(\lambda) = H_0 \left(\frac{\lambda}{\lambda_0} \right)^{-n} \quad \text{for } \lambda_1 < \lambda < \lambda_u \tag{8-1}$$

The dimensionless relaxation exponent n is allowed to take the values between 0 and 1. The front factor H_0, with the dimension Pa and the characteristic time λ_0, depends on the specific choice of material. Various values have been assigned in the literature. The spectrum has only two independent parameters, since several constants are lumped into $(H_0 \lambda_0^{-n})$. For certain materials (the special case of LST), the upper limit of the power law spectrum may diverge to infinity, $\lambda_u \to \infty$, without becoming inconsistent [18].

If the self-similar spectrum extends over a sufficiently wide time window, approximate solutions for the relaxation modulus $G(t)$ and the dynamic moduli $G'(\omega)$, $G''(\omega)$ might be explored by neglecting the end effects

$$G(t) - G_e = H_0 \int_0^\infty \left(\frac{\lambda}{\lambda_0} \right)^{-n} e^{-t/\lambda} \frac{d\lambda}{\lambda} = H_0 \Gamma(n) \left(\frac{t}{\lambda_0} \right)^{-n} \tag{8-2}$$

$$G'(\omega) = H_0 \Gamma(n) \Gamma(1 - n) \cos \frac{n\pi}{2} (\lambda_0 \omega)^n \tag{8-3}$$

$$G''(\omega) = H_0 \Gamma(n) \Gamma(1 - n) \sin \frac{n\pi}{2} (\lambda_0 \omega)^n \tag{8-4}$$

These solutions of the idealized problem are a good approximation for the behavior within a time window $\lambda_1 < t < \lambda_u$ or the corresponding frequency window $1/\lambda_u < \omega < 1/\lambda_1$. Truncation effects can be seen near the edges λ_1 and λ_u. λ_0 is some material-specific reference time, which has to be specified in each choice of material, and $H_0 \Gamma(n) = G_0$ is the corresponding modulus value.

The self-similar behavior is most obvious when it occurs in this form, i.e. if the exponent is negative and the self-similar region is extensive. $G(t)$, $G'(\omega)$, $G''(\omega)$, and $H(\lambda)$ all have power law format and they have been used interchangeably in the literature. Less obvious is the self-similar behavior for *positive* exponent values.

8.2 Self-Similar Relaxation with Positive Exponent Value

For positive exponent values, the symbol m with $m > 0$ is used. The spectrum has the same format as in Eq. 8-1, $H(\lambda) = H_0(\lambda/\lambda_0)^m$, however, the positive exponent results in a completely different behavior. One important difference is that the upper limit of the spectrum, λ_u, has to be finite in order to avoid divergence of the linear viscoelastic material functions. This prevents the use of approximate solutions of the above type, Eqs. 8-2 to 8-4.

Spectra with a positive exponent may be explored for the ideal case of power law relaxation over all times up to the longest relaxation time, λ_{max}:

$$H(\lambda) = H_0 \left(\frac{\lambda}{\lambda_{max}} \right)^m \quad \text{for } 0 < \lambda < \lambda_{max} \tag{8-5}$$

λ_{max} is always finite and is chosen here as the characteristic time of the spectrum. Even for this ideal spectrum, the relaxation modulus has to be evaluated numerically. It does not have any simple form which could be recognized as self-similar behavior. However, material functions can be evaluated for steady shear flow

$$\eta_0 = \int_0^{\lambda_{max}} H(\lambda) \, d\lambda = \frac{H_0 \lambda_{max}}{1 + m} \tag{8-6}$$

$$\psi_1 = 2 \int_0^{\lambda_{max}} H(\lambda) \lambda d\lambda = \frac{2 H_0 \lambda_{max}^2}{2 + m} \tag{8-7}$$

8.3 Observations of Self-Similar Relaxation Spectra

A wide variety of polymeric materials exhibit self-similar relaxation behavior with positive or negative relaxation exponents. Positive exponents are only found with highly entangled chains if the chains are linear, flexible, and of uniform length [61]; the power law spectrum here describes the relaxation behavior in the entanglement and flow region.

Power law relaxation behavior is also expected (or has already been found) for other critical systems. Even molten polymers with linear chains of high molecular weight relax in a self-similar pattern if all chains are of uniform length [61].

Self-similar spectra with negative exponents are found in several different systems such as microgels [155], polymer blends and block copolymers [156] at their critical point, or coagulating systems at the threshold. Some broadly distributed polymers exhibit power law relaxation over an intermediate frequency range [157]. This behavior, however, is not related to gelation, which would require the power law to extend into the terminal zone, $\omega \to 0$. The relaxation spectrum of a polymer is broadened as long range connectivity develops (divergence of longest relaxation time), resulting in a power law behavior with negative exponent. Indications can also be seen in branched polymers [158] and in solutions containing polymeric fractals (e.g. flexible chain macromolecules of arbitrary self-similar connectivity [159] or natural objects with non-integer dimension such as aggregates or percolation clusters [160]).

Various types of power law relaxation have been observed experimentally or predicted from models of molecular motion. Each of them is defined in its specific time window and for specific molecular structure and composition. Examples are dynamically induced glass transition [90, 161], phase separated block copolymers [162, 163], polymer melts with highly entangled linear molecules of uniform length [61, 62], and many others. A comprehensive review on power law relaxation has been recently given by Winter [164].

9 Applications

Processing of polymeric materials almost always involves a liquid-solid transition, and applications are often limited by solid-liquid transitions. Property changes are most dramatic near the transition. It is important to know where the transition occurs, how extensive the property changes are, and how fast they occur. Avoiding these transitions might often be the simplest solution, but that is not always an option. Some materials have to be produced, processed, or used near the gel point or up to the gel point. That is where rheological experiments permit the exact determination of the instant of gelation and the 'distance' from the gel point. We now can produce materials at controlled distances from the gel point and also process materials near the gel point. The critical gel properties serve as reference for expressing property changes in the vicinity of the gel point.

9.1 Avoiding the Gel Point

Mixing and shaping operations in polymer processing require sufficient molecular mobility which vanishes when the motion slows down near the gel point. Some materials only have small processing windows near the gel point (because of their limited chemical stability above their melting point, for instance, or because of their rapid crosslinking). Processing will become reasonably easy if such a narrow processing window can be targeted. This requires sufficiently accurate measurement of the material status relative to the location of the gel point. Instead of processing in the material characteristic window, one may consider shifting of the processing window by alteration of the material. This again requires detailed knowledge of the transition behavior and accurate methods for detecting the transition.

9.2 Materials Near the Gel Point

Controlled sample preparation is difficult near the gel point where the rate of property change is largest. Physical gelation usually proceeds too rapidly so that the material near the gel point eludes the experiment or the application. However, chemical gelation is most suitable for controlling the evolving network structure. Several approaches have been explored in industrial applications and in research laboratories:

1. Polymerization chemistry was developed which allows the stopping of the crosslinking reaction in the vicinity of the gel point [5, 29].
2. Thermal quenching of the crosslinking reaction: In an actual reactive extrusion process, the degree of crosslinking can be controlled by adjusting the residence time at elevated temperature.
3. Off-balancing of stoichiometry by the right amount (depending on crosslinking system) allows preparation of materials near the gel point [66].
4. Crosslinking by controlled amounts of radiation. Examples are γ-irradiated polyethylenes [165] and UV-irradiated polyurethane [166]. Best contestants are endothermic reactions which require energy for the formation of each chemical bond and which cease crosslinking when the energy supply is turned off. The extent of reaction directly depends on the amount of absorbed energy. Radiation can be used to enter energy into transparent materials. The radiation intensity decreases along the path of the radiation in the materials. This potentially leads to samples with a large gradient in extent of reaction, the exposed side of the sample being further crosslinked than the backside. Not suitable for preparing polymers near the gel point are chemical reactions in which radiation only initiates the crosslinking reaction so that it continues even after the radiation has been turned off.

9.3 Damping Materials

Critical gels have a damping plateau instead of the commonly observed damping peak. The loss tangent, Eq. 4-12, is uniformly high over a wide frequency range, $0 < \omega < 1/\lambda_0$. At higher frequencies, $\omega > 1/\lambda_0$, the usual glassy behavior sets in, or, if the critical gels are made from precursors of high molecular weight, the entanglement behavior dominates before glassy modes take over. The damping behavior is independent of temperature $[n \neq f(T)]$, which seems to be unique among polymeric materials.

The damping material does not have to be a critical gel. Many applications do not require extra low damping frequencies. The lowest vibration damping frequency ω_{min} determines the longest relaxation time, λ_{max}. A suitable damping material would be crosslinked beyond the gel point, with a λ_{max} of about $1/\omega_{min}$.

9.4 Pressure-Sensitive Adhesives

Polymers at the gel point are extremely powerful adhesives. They combine the surface-wetting property of liquids with the cohensive strength of solids. The mechanical strength of an adhesive bond in *composite materials* (with crosslinking matrix) develops during the sol-gel transition, and the strength of *pressure-sensitive adhesives* can be tailored through their degree of crosslinking. While the mechanical strength against adhesive failure is maximum at the gel point, the mechanical strength against cohesive failure is still relatively low since the polymer *at* the gel point is only slightly crosslinked. As the crosslink density is increased beyond the gel point, the strength of the network increases (stronger cohesion) while the adhesive strength decreases. The distance of a polymer from the gel point therefore is expected to define the ratio of adhesive to cohesive strength. This general behavior was confirmed by Zosel [167] in his study of tack and peel behavior of radiation-crosslinked PDMS. He found the maximum tack (corresponding to adhesive strength) slightly after the gel point.

The knowledge of gelation leads to an unconventional but systematic approach to the development of pressure-sensitive adhesives. Based on the fact that gelation is a critical phenomenon, we hypothesize that there exists a universal 'law' which relates the adhesive to the cohesive properties at the gel point and in its vicinity. More research is needed for verifying this hypothesis and exploring its limits. For that purpose, polymer gels with well-defined chemical composition should be prepared at known distance from the gel point and their adhesive and rheological properties measured. The main parameters are: solid surface properties (chemical composition, homogeneity, regularity, roughness, curvature), chemical composition, wetting (phase diagram), and layer thickness.

The molecular weight is one of the most important parameters of polymer adhesion. Molecules of low molecular weight are able to wet a surface without adhering there for a longer period of time. There exists a chemistry-specific molecular weight beyond which molecules adhere to a surface. The transition from non-adhering to adhering may be described as a critical phenomenon. This same phenomenon has important implications for gelation near a surface: The molecular weight increases during gelation, and, at a certain extent of crosslinking, the largest cluster exceeds the critical molecular weight and adheres to the surface. This may have interesting consequences which should be explored further. The largest clusters may separate from the bulk, and the gel point will then be postponed in a very thin film while a solid layer is formed near the surface. Beyond the gel point, large molecules will not be able to move to the surface and the system will gradually feel 'dry', i.e. will not be able to stick to a surface.

Molecular orientation at the surface may also be important. A molecule orients planarly when deposited on a solid surface. Molecular strands prefer to be parallel to the surface; their probability of being oriented normal to the surface is very low. Several mechanisms can cause this orientation: (1) Surface-active sites may favor entire chain segments to interact with the surface. (2) The orientation is generated during the drying of the solution at the surface. A volume element of solution at the surface may initially contain isotropic polymer chains. Extraction of the solvent results in a shrinkage of the volume element into a thin film at the surface. Motion of the chain segments during this 'deformation' results in planar orientation.

9.5 Processing Near the Physical Gel Point

Many polymers solidify into a semi-crystalline morphology. Their crystallization process, driven by thermodynamic forces, is hindered due to entanglements of the macromolecules, and the crystallization kinetics is restricted by the polymer's molecular diffusion. Therefore, crystalline lamellae and amorphous regions coexist in semi-crystalline polymers. The formation of crystals during the crystallization process results in a decrease of molecular mobility, since the crystalline regions act as crosslinks which connect the molecules into a sample spanning network.

Injection molding of semi-crystalline polymers is an example of a process which is dominated by crystallization. Particularly the rate of crystallization is important. During injection of the molten polymer into the mold, a layer of polymer gets deposited at the cold walls of the mold where it starts solidifying while exposed to high shear stress due to ongoing injection (Fig. 27). This is an important part of the process, since the wall region later forms the surface of the manufactured product. A high shear stress during this crystallization might rupture the already solidified surface layers, resulting in surface defects (Y.G. Lin, personal communication). This could be avoided if the crystallization behavior were known together with its effect on the developing strength of the material. The polymer could to be modified to adjust its crystallization behavior

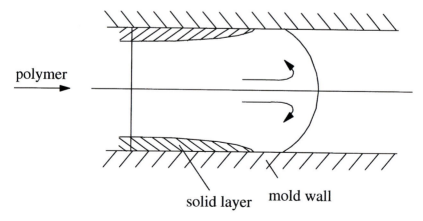

polymer

solid layer   mold wall

Fig. 27. Schematic of molten polymer flow during injection molding into a cold mold

or a different polymer could be used or developed. A suitable tool to monitor the time dependence of crystallization processes and thus to select the best material and processing conditions is rheology. The characteristic crystallization time for each material can be tailored to the given molding circumstances.

9.6 Processing Near the Chemical Gel Point

Novel materials can be produced by crosslinking a polymer while shearing it at the same time. This process is known as dynamic vulcanization and is already used in the manufacture of certain rubbers and elastomers. At low degrees of crosslinking, the molecular structure is still very soft, so that the rate of molecular breaking can be of the same order as the rate of forming new chemical bonds. The result is a highly branched molecular structure with molecules of broadly distributed size. The materials have advantageous properties for applications as adhesives, damping materials with extremely broad damping maxima, toner for copy machines, superabsorbers, and sealants. The process gives access to materials with a wide range of properties through well-defined processing but to uniform chemical composition (mono-material, easy to recycle).

10 Conclusions

Polymers during their liquid-solid transitions develop a universal rheological behavior which is distinct from that of liquids or solids. It seems that this is a new, general phenomenon of nature. The phenomenological picture of this relaxation pattern is fairly complete. This allows the direct measurement of the liquid-solid transition during the manufacturing of gels and suggests a framework for presenting experimental data on gels. However, much too little is known yet about the molecular origin of the observed phenomena. Such molecular understanding would be desirable from a fundamental point of view, and it would be good to have when designing gels for specific applications.

We have further evidence that, in the absence of competing phase transitions, the material at the liquid-solid transition exhibits self-similar rheological behavior. The self-similarity expresses itself in power law relaxation and retardation spectra in the terminal zone. This very distinct relaxation pattern was first detected for chemically gelling systems. For these systems, a three-dimensional network structure is built by permanently connecting molecules through covalent bonds, and the longest relaxation time diverges at the gel point. However, the power law relaxation behavior seems to also govern physical gelation as long as the gel number is large, $N_g \gg 1$, i.e. the lifetime of the junctions (which are reversible in contrast to chemical gelation) is long compared to the experimental observation time. This can be clearly seen at early stages of crystallization where a molecular network is formed with crystalline junctions. Furthermore, other physical mechanisms which lead to increasing correlation length seem to result in the same general relaxation patterns near the liquid/solid transition point.

The dynamic properties depend strongly on the material composition and structure. This is not included in current theories, which seem much too ideal in view of the complexity of the experimentally found relaxation patterns. Experimental studies involving concurrent determination of the static exponents, d_f and τ, and the dynamic exponent, n, are required to find limiting situations to which one of the theories might apply.

The criticality of the liquid-solid transition does not necessarily imply universality for all investigated properties. Prefactors of different properties always depend on the details of the underlying structure, e.g. the critical gel stiffness is seen to vary by five orders of magnitude. Scaling of different static properties around a critical point is only universal if the structure growth process belongs to the same class (universality class). Dynamic scaling properties such as the power law relaxation exponent are much more complex and not expected to be universal. Values of n as reported in this study, $0.19 < n < 0.92$, have been established over almost the entire possible range $(0 < n < 1)$. Stoichiometry, molecular weight, and concentration were all shown to have an impact on the critical gel properties. These critical gel properties are also coupled for the polymers studied here.

From a practical point of view, it is advantageous that critical gel properties depend on molecular parameters. It allows us to prepare materials near the gel point with a wide range of properties for applications such as adhesives, absorbents, vibration dampers, sealants, membranes, and others. By proper molecular design, it will be possible to tailor network structures, relaxation character, and the stiffness of gels to one's requirements.

Acknowledgements We gratefully acknowledge financial support from CUMIRP, MRSEC, and the National Science Foundation under grants NSF-MSM-86-01595 and NSF-DMR-88-06556. One of us, M.M., is indebted to the German Academic Exchange Service for financial support (DAAD-Doktorandenstipendium aus Mitteln des zweiten Hochschulsonderprogramms). We would also like to thank K. Dušek and K. te Nijenhuis for critically reading this text and for their helpful comments and suggestions.

11 References

1. Flory PJ (1941) J Am Chem Soc 63: 3083
2. Stockmayer WH (1943) J Chem Phys 11: 45
3. Stauffer D (1985) Introduction to percolation theory. Taylor and Francis, Philadelphia, USA
4. Vilgis TA, Winter HH (1988) Coll Polym Sci 266: 494
5. Chambon F, Winter HH (1985) Polym Bull 13: 499
6. Winter HH, Chambon F (1986) J Rheol 30: 367
7. Chambon F, Winter HH (1987) J Rheol 31: 683

8. Baumgärtel M, Winter HH (1989) Rheol Acta 28: 511
9. Chasset R, Thirion P (1965) In: JA Prins (ed) Physics of non-crystalline solids. North-Holland, Amsterdam
10. Ferry JD (1980) Viscoelastic properties of polymers. Wiley, New York
11. Pipkin AC (1986) Lectures on viscoelasticity theory, 2nd edn. Springer, Berlin Heidelberg, New York
12. Orbey N, Dealy JM (1991) J Rheol 35: 1035
13. Martin JE, Adolf D, Wilcoxon JP (1989) Phys Rev A 39: 1325
14. Friedrich C, Heymann L (1988) J Rheol 32: 235; Friedrich C, Heymann L, Berger HR (1989) Rheol Acta 28: 535
15. Adolf D, Martin JE (1990) Macromolecules 23: 3700
16. Stauffer D, Coniglio A, Adam A (1982) Adv Polym Sci 44: 103
17. Winter HH, Izuka A, De Rosa ME (1994) Polym Gels Networks 2: 239
18. Winter HH (1987) Progr Colloid Polym Sci 75: 104
19. Winter HH (1991) MRS Bulletin 16(8): 44
20. Goldbart P, Goldenfeld N (1992) Phys Rev A 45: R5343
21. Vilgis TA (1992) Polymer Networks - Crosslinking, Rubbers, Microgels Ch. 33
22. Flory PJ (1953) Principles of polymer chemistry. Cornell University Press, Ithaca, New York
23. Miller DR, Macosko CW (1976) Macromolecules 9: 206
24. Miller DR, Vallés EM, Macosko CW (1979) Polym Eng Sci 19: 272
25. Dusek K, Ilavsky M (1974) J Polym Sci C 53: 57
26. Stanford JL, Stepto RFT, Waywell DR (1975) Faraday Disc 71: 1308
27. Gordon M, Scantlebury GR (1996) Proc Roy Soc London A292: 380
28. Gordon M, Templ WB (1972) Makromol Chem 160: 263
29. Vallés EM, Macosko CW (1979) Macromolecules 12: 521
30. Venkataraman SK, Coyne L, Chambon F, Gottlieb M, Winter HH (1989) Polymer 30: 2222
31. De Rosa ME (1994) Dissertation. University of Massachusetts at Amherst
32. Broadbent SR, Hammersley JM (1957) Proc Cambridge Phil Soc 53: 629
33. de Gennes PG (1979) Scaling concepts in polymer physics. Cornell, Ithaca, New York
34. Martin JE, Adolf D (1991) Annu Rev Phys Chem 42: 311
35. Colby RH, Rubinstein M, Gillmor JR, Mourey TH (1992) Macromolecules 25: 7180
36. Schosseler S, Leibler L (1984) Physique Lett 45: 501
37. Stauffer D (1981) Pure Appl Chem 53: 1479
38. Colby RH, Gillmor JR, Rubinstein M (1993) Phys Rev E 48: 3712
39. Martin JE, Adolf D, Wilcoxon JP (1988) Phys Rev Lett 61: 2620
40. Daoud M (1988) J Phys A 21: L237
41. de Gennes PG (1977) J Physique Lett 38: L-355
42. Muthukumar M, Winter HH (1986) Macromolecules 19: 1284
43. Hess W, Vilgis TA, Winter HH (1988) Macromolecules 21: 2536
44. Muthukumar M (1989) Macromolecules 22: 4656
45. Smoluchowski MV (1916) Phys Z 17: 585
46. Drake RL (1972) In: GM Hidy, JR Brock (ed) Topics in current aerosol research, vol 3. Pergamon, New York
47. Ernst MH (1984) In: EGD Cohen (ed) Fundamental problems in statistical mechanisms, vol. VI North-Holland, Amsterdam
48. van Dongen PGJ, Ernst MH (1985) Phys Rev Lett 54: 1396
49. Hendricks EM, Ernst MH, Ziff RM (1983) J Stat Phys 31: 519
50. Ziff RM, Ernst MH, Hendricks EM (1983) J Phys A 16: 2293
51. Leung YK, Eichinger BE (1984) J Chem Phys 80: 3877; (1984) J Chem Phys 80: 3885
52. Shy LY, Eichinger BE (1985) Brit Polym J 17: 200
53. Shy LY, Leung YK, Eichinger BE (1985) Macromolecules 18: 983
54. Bird RB, Armstrong RW, Hassager O (1987) Dynamics of Polymerics Liquids, vol 1. Wiley, New York
55. Lodge AS (1964) Elastic liquids: an introductory vector treatment of finite-strain polymer rheology. Academic, London
56. Chang H, Lodge AS (1972) Rheol Acta 11: 127
57. Larson R (1987) Constitutive equations for polymer melts and solutions. Butterworth, London
58. Scanlan JC, Winter HH (1991) Makrom Chem, Makrom Symp 45: 11
59. Izuka A, Winter HH, Hashimoto T (1992) Macromolecules 25: 2422

60. De Rosa ME, Winter HH (1994) Rheol Acta 33: 220
61. Baumgärtel M, Schausberger A, Winter HH (1990) Rheol Acta 29: 400
62. Baumgärtel M, De Rosa ME, Machado J, Masse M, Winter HH (1992) Rheol Acta 31: 75
63. Hogdson DF, Amis EJ (1990) Macromolecules 23: 2512
64. Lairez D, Adam M, Emery JR, Durand D (1992) Macromolecules 25: 286
65. Muller R, Gérard E, Dugand P, Rempp P, Gnanou Y (1991) Macromolecules 24: 1321
66. Durand D, Delsanti M, Adam M, Luck JM (1987) Europhys Lett 3: 297
67. te Nijenhuis K, Winter HH (1989) Macromolecules 22: 411
68. Lin YG, Mallin DT, Chien JCW, Winter HH (1991) Macromolecules 24: 850
69. Richtering HW, Gagnon KD, Lenz RW, Fuller RC, Winter HH (1992) Macromolecules 25: 2429
70. Mours M, Winter HH (1996) Macromolecules 29: 7221
71. Venkataraman SK, Winter HH (1990) Rheol Acta 29: 423
72. Cole KS, Cole RH (1941) J Chem Phys 9: 341
73. Gross B (1953) Mathematical structures of the theory of viscoelasticity. Hermann, Paris
74. Cates ME (1987) Macromolecules 20: 2289
75. Cates ME, Candau SJ (1990) J Phys: Condens Matter 2: 6869
76. te Nijenhuis K (1995) Viscoelastic properties of thermoreversible gels, submitted to Adv Polym Sci
77. Keller A (1995) Faraday Disc 101: 1
78. Lehsaini N, Muller R, Weill G, François J (1994) Polymer 35: 2180
79. Yu Q, Amis EJ (1993) Makromol Chem, Makromol Symp 76: 193
80. Schwittay C, Mours M, Winter HH (1995) Faraday Disc 101: 93
81. Guenet JM (1992) Thermoreversible gelation of polymers and biopolymers. Academic, London
82. Reid DS, Bryce TA, Clark AH, Rees DA (1974) Faraday Disc 57: 230
83. Ross-Murphy SB (1991) Rheol Acta 30: 401
84. Michon C, Cuvelier G, Launay B (1993) Rheol Acta 32: 94
85. Wedler W, Tang W, Winter HH, MacKnight WJ, Farris RJ (1995) Macromolecules 28: 512
86. Winter HH, Scott DB, Gronski W, Okamoto S, Hashimoto T (1993) Macromolecules 26: 7236
87. Larson RG, Winey KI, Patel SS, Watanabe H, Bruinsma R (1993) Rheol Acta 32: 245
88. Vasilev VG, Rogovina LZ, Slonimskii GL, Papkov VS, Shchegolikhina OI, Zhdanov AA (1995) Polym Sci, Ser A 37: 174
89. Rogovina LZ, Vasilev VG, Papkov VS, Shchegolikhina OI, Slonimskii GL, Zhdanov AA (1995) Macromol Symp 93: 135
90. Tobolsky AV, McLouhglin JR (1952) J Polym Sci 8: 543
91. Cheng YT, Johnson WL (1987) Science 235: 997
92. Zorn R, McKenna GB, Willner L, Richter D (1995) Macromolecules 28: 8552
93. Osaki K, Inoue T, Hwang EJ, Okamoto H, Takiguchi O (1994) J Non-cryst Solids 172–174: 838
94. Colby RH, Gillmor JR, Galli G, Laus M, Ober CK, Hall E (1993) Liquid Cryst 13: 233
95. Rubin SF, Kannan RM, Kornfield JA, Boeffel C (1994) Proc ACS Polym Mater Sci Eng 71: 330, 486
96. Münstedt H (1975) Angew Makrom Chem 47: 229
97. Zosel A (1972) Rheol Acta 11: 229
98. Laun HM (1984a) Angew Makromol Chem 123/124: 335
99. White JL (1979) J Non-Newt Fluid Mech 5: 177
100. Buscall R, Goodwin JW, Hawkins MW, Ottewill RH (1982) J Chem Soc Faraday Trans I 78: 2873, 2889
101. Münstedt H (1981) Polym Eng Sci 21: 259
102. Castellani L, Lomellini P (1991) Plastics, Rubber and Composites Processing and Applications 16: 25
103. Minagawa N, White JL (1975) Polym Eng Sci 15: 825
104. Krieger IM (1972) Adv Colloid Interface Sci 3: 111
105. Krieger IM, Eguiluz M (1976) Trans Soc Rheol 20: 29
106. Hoffmann RL (1972) Trans Soc Rheol 16: 155
107. Hoffmann RL (1974) J Colloid Interface Sci 46: 491
108. Jeffrey DJ, Acrivos A (1976) AIChE J 22: 417
109. Russel WB, Saville DA, Schowalter WR (1989) Colloidal Dispersions. Cambridge University Press, Cambridge

110. Laun HM (1984) Colloid Polym Sci 262: 257
111. Metzner AB (1985) J Rheol 29: 739
112. Masuda T, Nakajima A, Kitamura M, Aoki Y, Yamauchi N, Yoshioka A (1984) Pure Appl Chem 56: 1457
113. Weinberger CB, Goddard JD (1974) Intern J Multiphase Flow 1: 465
114. Castellani L, Lomellini P (1994) Rheol Acta 33: 446
116. Baumgärtel M, Winter HH (1992) J Non-Newt Fluid Mech 44: 15
115. Venkataraman SK (unpublished results)
117. Mours M, Winter HH (1994) Rheol Acta 33: 385
118. te Nijenhuis K, Dijkstra D (1975) Rheol Acta 14: 71
119. Winter HH, Morganelli P, Chambon F (1988) Macromolecules 21: 532
120. Holly EE, Venkataraman SK, Chambon F, Winter HH (1988) J Non-Newt Fluid Mech 27: 17
121. Izuka A, Winter HH, Hashimoto T (1994) Macromolecules 27: 6883
122. Scanlan JC, Winter HH (1991) Macromolecules 24: 47
123. Flory PJ (1974) Faraday Disc 57: 7
124. Schmidt M, Burchard W (1981) Macromolecules 14: 370
125. Martin JE, Wilcoxon JP, Adolf D (1987) Phys Rev A 36: 1803
126. Martin JE, Wilcoxon JP (1988) Phys Rev Lett 61: 373
127. Martin JE, Keefer KD (1986) Phys Rev A 34: 4988
128. Kajiwara K, Burchard W, Kowalski M, Nerger D, Dusek K, Mateijka L, Tuzar Z (1984) Makromol Chem 185: 2543
129. Vallés EM, Macosko CW (1979) Macromolecules 12: 673
130. Allain C, Amiel C (1986) Phys Rev Lett 56: 1501
131. Barton JM, Buist GJ, Hamerton I, Howlin BJ, Jones RJ, Lin S (1994) Polym Bull 33: 215
132. Mijovic J, Kim J, Slaby J (1984) J Appl Polym Sci 29: 1449
133. Muzumdar SV, Lee LJ (1991) Polym Eng Sci 31: 1647
134. Lee JH, Lee JW (1994) Polym Eng Sci 34: 742
135. Soltero J, González-Romero V (1988) Proc Ann Techn Meet Soc Plast Eng, ANTEC, pp 1057–1061
136. Walton JR, Williams KPJ (1991) Vibr Spectr 1: 339
137. Lyon RE, Chike KE, Angel SM (1994) J Appl Polym Sci 53: 1805
138. Macosko CW (1985) Brit Polym J 17: 239
139. Lipshitz S, Macosko CW (1976) Polym Eng Sci 16: 803
140. Vallés EM, Macosko CW (1976) Rubber Chem Tech 49: 1232
141. Castro JM, Macosko CW, Perry SJ (1984) Polym Commun 25: 82
142. Apicella A, Masi P, Nicolais L (1984) Rheol Acta 23: 291
143. Adam M, Delsanti M, Durand D (1985) Macromolecules 18: 2285
144. Malkin AY (1985) Plaste Kautschuk 32: 281
145. Bidstrup SA (1986) Dissertation University of Minnesota
146. Allain C, Salomé L (1987) Polym Commun 28: 109
147. Axelos MAV, Kolb M (1990) Phys Rev Lett 64: 1457
148. Farris RJ, Lee C (1983) Polym Eng Sci 23: 586
149. Winter HH (1987) Polym Eng Sci 27: 1698
150. Cuvellier G, Peighy-Nourry C, Launay B (1990) In: S Phillips (ed) Gums and stabilizers for the food industry. IRL, Oxford, UK, pp 549–552
151. Malkin AY, Kulichikhin SG (1991) Adv Polym Sci 101: 217
152. Gillham JK (1976) Adv Polym Sci 19: 319; (1976) Adv Polym Sci 19: 676
153. ASTM Standard (1985) D4473–85
154. Tung CYM, Dynes PJ (1982) J Appl Polym Sci 27: 569
155. Antonietti M, Fölsch KF, Sillescu H, Pakula T (1989) Macromolecules 22: 2812
156. Bates FS (1984) Macromolecules 17: 2607
157. Larson R (1985) Rheol Acta 24: 327
158. Roovers J, Graessley WW (1981) Macromolecules 14: 766
159. Cates ME (1985) J Physique 46: 1059
160. Muthukumar M (1985) J Chem Phys 83: 3161
161. Andrews RD, Tobolsky AV (1951) J Polym Sci 7: 221
162. Gouinlock EV, Porter RS (1977) Polym Eng Sci 17: 535
163. Chung CI, Lin MI (1978) J Polym Sci, Polym Phys Ed 16: 545

164. Winter HH (1994) J Non-cryst solids 172–174: 1158
165. Vallés EM, Carella JM, Winter HH, Baumgärtel M (1990) Rheol Acta 29: 535
166. Khan SA, Plitz IM, Frank RA (1992) Rheol Acta 31: 151
167. Zosel A (1991) J Adhesion 34: 201

Editor: Prof. K. Dusek
Received: October 1996

Author Index Volumes 101-134

Author Index Volumes 1-100 see Vol. 100

Subject Index

Printing: Saladruck, Berlin
Binding: Buchbinderei Lüderitz & Bauer, Berlin